SMART DESIGN, SCIENCE AND TECHNOLOGY

Smart Science, Design and Technology

ISSN: 2640-5504
eISSN: 2640-5512

Book Series Editors

Stephen D. Prior
Faculty of Engineering and Physical Sciences
University of Southampton
Southampton
UK

Siu-Tsen Shen
Department of Multimedia Design
National Formosa University
Taiwan, R.O.C.

VOLUME 4

PROCEEDINGS OF THE IEEE 6TH INTERNATIONAL CONFERENCE ON APPLIED SYSTEM INNOVATION (ICASI 2020), 5–8 NOVEMBER 2020, TAITUNG, TAIWAN

Smart Design, Science and Technology

Editors

Artde Donald Kin-Tak Lam
Fujian University of Technology, P.R. China

Stephen D. Prior
University of Southampton, Southampton, UK

Sheng-Joue Young
National Formosa University, Huwei District, Taiwan

Siu-Tsen Shen
National Formosa University, Huwei District, Taiwan

Liang-Wen Ji
National Formosa University, Huwei District, Taiwan

CRC Press
Taylor & Francis Group
Boca Raton   London   New York   Leiden

CRC Press is an imprint of the
Taylor & Francis Group, an **informa** business

A BALKEMA BOOK

CRC Press/Balkema is an imprint of the Taylor & Francis Group, an informa business

© 2021 Selection and editorial matter, the Editors; individual chapters, the contributors

Typeset by MPS Limited, Chennai, India

Library of Congress Cataloging-in-Publication Data
A catalog record has been requested for this book

Published by: CRC Press/Balkema
                Schipholweg 107C, 2316 XC Leiden, The Netherlands
                e-mail: enquiries@taylorandfrancis.com
                www.routledge.com – www.taylorandfrancis.com

ISBN: 978-1-032-01993-2 (HBk)
ISBN: 978-1-032-03688-5 (Pbk)
ISBN: 978-1-003-18851-3 (eBook)
DOI: 10.1201/9781003188513

Smart Design, Science and Technology – Lam et al (eds)
© 2021 the Author(s), ISBN 978-1-032-01993-2

Table of contents

Preface

We have great pleasure in presenting this conference proceeding for technology applications in engineering science and mechanics from the selected articles of the International Conference on Applied System Innovation (ICASI 2020), organized by the International (Taiwanese) Institute of Knowledge Innovation and the IEEE, 5-8 November, 2020 at Taitung, Taiwan.

The ICASI 2020 conference was a forum that brought together users, manufacturers, designers, and researchers involved in the structures or structural components manufactured using smart science. The forum provided an opportunity for exchange of the research and insights from scientists and scholars thereby promoting research, development and use of computational science and materials. The conference theme for ICASI 2020 was "Innovation in a Post-COVID World" and tried to explore the important role of innovation in the development of the technology applications, including articles dealing with design, research, and development studies, experimental investigations, theoretical analysis and fabrication techniques relevant to the application of technology in various assemblies, ranging from individual to components to complete structure were presented at the conference. The major themes on technology included Material Science & Engineering, Communication Science & Engineering, Computer Science & Engineering, Electrical & Electronic Engineering, Mechanical & Automation Engineering, Architecture Engineering, IOT Technology, and Innovation Design. About 200 participants representing 11 countries came together for the 2020 conference and made it a highly successful event. We would like to thank all those who directly or indirectly contributed to the organization of the conference.

Selected articles presented at the ICASI 2020 conference will be published as a series of special issues in various journals. In this conference proceeding we have some selective articles from various themes. A committee consisting of experts from leading academic institutions, laboratories, and industrial research centres was formed to shortlist and review the articles. The articles in this conference proceeding have been peer reviewed to the usual standards. We are extremely happy to bring out this conference proceeding and dedicate it to all those who have made their best efforts to contribute to this publication.

Professor Siu-Tsen Shen & Dr Stephen D. Prior

Smart Design, Science and Technology – Lam et al (eds)
© 2021 the Author(s), ISBN 978-1-032-01993-2

Editorial Board

Smart Design, Science and Technology – Lam et al (eds)
© 2021 the Author(s), ISBN 978-1-032-01993-2

The development of augmented reality learning applications for mobile devices- an example of a personal portfolio

Chun-Ping Wu
Associate Professor, Department of Education, National University of Tainan, Tainan City, Taiwan, ROC

Chien-Yu Kuo*
Department of Applied Physics, National University of Kaohsiung, Kaohsiung, Taiwan

ABSTRACT: This study is to develop AR (Augmented Reality) materials for M-learning (mobile-learning) built by the website development tools Vuforia and Unity. Not only are these for server application across various platforms (Windows, Mac-OS), but they also support teachers in personal portfolios to prepare the materials dependent on the varied teaching sites. This refers to the facility, layout, system plan and the teaching environment of different M-learning devices, such as laptop/tablet/smart phone for the OS (Operating System) of Windows, iOS & Android in a classroom of a general primary school.

The teaching materials in this study had been tested by the subjects, including Windows, iOS and Android operation systems of the multiple M-learning devices (laptop, tablet, smartphones), which are in the teaching environments (a classroom in the elementary school, teacher's office and university campus). The sequence of these tests in this study is "laptop-tablet-smartphone," which requires the mouse device in laptop tests to support the subjects to learn about the test process and doesn't require any supported device in the other two tests.

This study builds the AR materials for M-learning to be used in a local mobile platform in cloud storage for teachers. It is easily used in sharing and follow-up development. This process of building the platform can be referred for teachers from different fields.

1 INTRODUCTION

The rapid development of science and technology not only facilitates the dissemination of information, but also makes the information situation more diversified. Therefore, different educational methods have emerged. More and more tablet computers are used to assist teaching. Today, society is constantly advancing, technology is constantly evolving and more and more digital media is appearing. This situation is not only because handheld devices have gradually integrated into our daily lives, but also because of the development of the Internet since even information is spreading both slowly and quickly. In many cases, only handheld devices can search for the information you need. It is equivalent to a small computer that can be carried around. It has become an indispensable tool and source of entertainment in today's society.

Traditional teaching materials often use books, pictures or audiovisual teachings to convey knowledge. This static teaching method is unlikely to catch a digital native's attention, nor is it conducive to conveying the comprehensiveness of knowledge. George Miller, a modern psychologist, discovered that short-term

memory can only hold 7 (plus or minus two) things at once. The information here is also stored for only 15–20 seconds. The information stored in the short-term memory can be committed to the long-term memory store. There is no limit for the information stored in the long-term memory. The information stored here can stay for many years. It is mentioned in the psychology IP theory that long-term memory can be divided between the semantic, episodic and procedural memories. The semantic memory is made up of facts or information learned or obtained throughout the life. The episodic memory is made up of personal experiences or real events that have happened in a person's life. Lastly, the procedural memory is made up of procedures or processes learned such as riding a bike (Wikipedia). Therefore, we can know that if we want to turn the received knowledge into long-term memory, it is more than just learning with pictures or words. Practical, operation or situational learning accounts for three-halves of long-term memory.

On the other hand, in dual-coding theory, a theory of cognition, it is mentioned that only words and images are used in mental representation. But if a relevant visual is also presented or if the learner can imagine a visual image to go with the verbal information, that memory for some verbal information is enhanced. Likewise, visual information can often be

---

*Corresponding Author

enhanced when paired with relevant verbal information, whether real-world or imagined. This theory has been applied to the use of multimedia presentations. Because multimedia presentations require both spatial and verbal working memory, individuals dual-code information presented and are more likely to recall the information when tested at a later date. Because multimedia presentations require both spatial and verbal working memory, individuals dual code information presented and are more likely to recall the information when tested at a later date. Moreover, studies that have been conducted on abstract and concrete words have also found that the participants remembered concrete words better than the abstract words. (Wikipedia) Therefore, AR technology in the past is better than traditional teaching methods because it uses the traditional expression of TV projectors, combined with modern portable dining table equipment. By comparing the lens of a mobile device application with a special drawing card, you can display 3D sound animation, a more comprehensive explanation of knowledge points and more interactive content. Vivid images can enhance students' memory points. It also has extremely high portability, forming a simple and efficient learning method.

Traditional teaching is the "injection method" and also the "explanation method". It is made from a combination of speech and presentation. In this way, the teacher speaks and the students listen. The purpose is to explain the writing method to students, or explain the nouns in the text; or describe the life of the author; or analyze the characters in the book to explore certain situations. Or determine the theories of various schools and instill knowledge and ideas. This is a passive process of imparting knowledge and cannot cultivate the motivation of students to explore things automatically. AR is a way to turn passive listening into active making. Edgar Dale (April 27, 1900 – March 8, 1985), a U.S. educationist, proposed the learning experience pyramid theory. Dale's "cone of experience" is a visual model consisting of eleven stages. The arrangement in the cone is based on abstraction and the number of senses involved. Experiences at each stage can be mixed and related to each other to promote more meaningful learning

The times are advancing. Facing new sentence patterns, there are many uncertainties in the teaching field, but it also offers many possibilities. Ideas never thought of before can now be realized. In the field of teaching, teachers can start to learn how to use new media in teaching. (Edgar Dell, experience in 1946) In addition, many high-end technology products have appeared in recent years. These include virtual reality and augmented reality. These names are "real-virtuality continuum" (real-virtuality continuum), proposed by Paul Milgram in 1994, it is the most widely used type of theoretical concept.

So it is the necessity of the preservice teacher to have the ability of making an AR. In addition to having the aforementioned advantages, it is easy to attract attention, provide better practical experience and stimulate student interest, etc. This research

focuses on the preservice teacher. You can cultivate practical experience by making a Vuforia of AR.

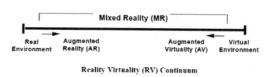

Picture 1.   Real-virtual continuum
Image source: Science and Technology Policy Research and Information Center (2016)
https://portal.stpi.narl.org.tw/index/article/10258.

2   AR FOR M-LEARNING

2.1   Mobile learning

Mobile learning (M-Learning) is a learning method that transcends geographical restrictions and makes full use of portable technology. In other words, action learning eliminates geographical restrictions. Action learning has different meanings in different social groups. Although it is related to online learning and distance learning, its obvious difference lies in comprehensive learning and learning using handheld devices.

As far as the term M-Learning is concerned, it includes the use of portable technology for learning, which focuses on this technology (it does not require a stable environment such as in a classroom); comprehensive learning focuses on the learners' mobility and interaction with portable or fixed technology. While learning in a mobile society, the focus is on how the whole society and its systems accept and support the learning of the ever-increasing mobile population. And it is no longer satisfied with the existing learning methodology.

The convenience of action learning is that it can be used anywhere and everywhere (classroom, taxi, laundromat, bathroom, etc.) as long as there are available learning apps. In addition, it is cooperative in nature, that is to say, almost everyone can use the same content immediately, and it will also bring immediate feedback which, in turn, will produce the best learning method.

Action learning replaces books and notes with a small memory full of streamlined learning content, which also brings great portability. In addition, this form of learning is full of charm and fun. With such a learning model, it is no longer difficult to combine games and learning more efficiently and gain more entertainment experience.

2.2   AR

AR is translated as "Augmented Reality", or AR for short. It is a technology that adds virtual objects and scenes to the real world. For example, "Pokémon GO" launched by Niantic in 2016 sparked a global trend. It also made many people realized the potential in

AR technology. This technology has prompted many manufacturers to take advantage of the opportunity to launch many applications related to AR.

2.3 AR is best for M-learning

AR is best for M-learning. Because after designing the content, you can easily put it into your mobile device and carry it everywhere for actual operation. In the process, you don't need a network connection. It simply identifies the graphics card and the mobile device can quickly present teaching content.

3 DESIGN

The process of augmented reality textbook design. This creation uses empirical study to conduct a case study of a single work and uses the Observational Method to conduct mobile learning analysis. In the early stage, the development framework process was drawn up (see Figure 2), mainly using Vuforia to identify augmented reality elements, supplemented by Unity interactive software, and importing audio and menu components for interface design. All relevant model materials used in this work are for self-creation or recording.

4 RESULTS AND DISCUSSION

The main purpose of this creation is to explore the combination of traditional works and augmented reality, and whether the display content can increase the willingness to learn, expand people's perceptual knowledge, inspire people's thinking and lead to new discoveries.

According to the questions raised by the research motivation of this work, through observational analysis, we can get the following conclusions:

1. We can learn that the use of handheld devices can solve the difficulty of learning, the difficulty of carrying multimedia hardware devices, field problems, and the increase of willingness to learn for students.
2. This analysis found that most of the students who presented immersive learning agreed. The handheld device not only solved the difficulty of learning, but also improved the willingness to learn, and immersed the subjects in it.

According to the actual test, I invited five first-year university students, two second-year university students, and two university fourth-year students. They are preservice teachers and conducted actual tests. Nearly 88% of them, regardless of contact or production AR, all for question 3, It is easy to learn to watch this AR portfolio with a mobile phone. Question 4: It is very easy to use the mobile phone to watch this AR portfolio. Question 5: I can use my mobile phone to watch this AR portfolio without much effort. Question 6: The various interactive functions in this AR portfolio

are clear and easy to use. Four questions express high agreement. After instructing the operation skills once, they can quickly get started and show great interest in this way of presentation. From this, we can start from question 7: Presenting the portfolio in AR can enhance the texture of the portfolio. Question 8: I think readers will prefer this type of AR portfolio. Question 9: I feel that using AR to present my portfolio can better demonstrate my abilities and strengths. Question 10: I like to use AR to present the concept of portfolios. But in question 11 ("If there is an opportunity in the future, I will want to learn how to make an AR portfolio," 22.2% of the subjects have low interest to try to make an AR portfolio. All of them are afraid of the unknown because they don't know enough about the AR production process. After the explanation, the acceptance has improved. Finally, question 12: I would like to promote the concept of AR portfolio to my friends. Question 13: Overall, the experience of using mobile phones to watch AR's portfolio is satisfactory. It is obvious that 77.8% of the subjects are satisfied with the viewing experience and willing to promote it to Friends.

Figure 1.   The trend of the distribution of testees by grade.

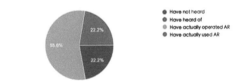

Figure 2.   The subject's AR experience.

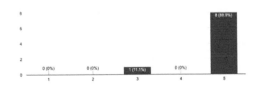

Figure 3.   It's easy to learn to watch AR works on your mobile phone.

3

4. It's easy to watch AR works using mobile phones

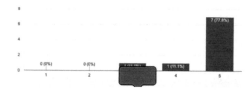

Figure 4. It's easy to watch AR works using mobile phones.

5. You can watch AR works on your mobile phone without much effort

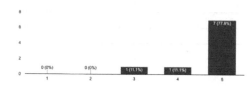

Figure 5. The subject's feelings about the degree of manipulation of the work.

6. The interactive functions in AR works are clear and easy to operate

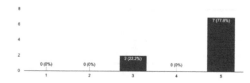

Figure 6. The subject's feelings about the ease of operation of the work.

7. The testees' feelings about the use of AR in their works and the texture of the works

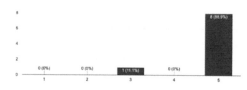

Figure 7. The testees' feelings about the use of AR in their works and the texture of the works.

5 CONCLUSION

In a report in Foresight Magazine in 2017, it was mentioned that 30% is learning and 70% is creation. For example, if students in the Department of Mechanical Engineering only learn how to assemble machinery from books, when they are actually implemented, there are actually more than half of them that still cannot understand how to do it, and even need to explore for a long time. Through AR augmented reality technology, they can better understand the assembly direction and correct position, and can avoid the loss of many materials.

REFERENCES

[1] Wang Yayi, 2020, Understanding Augmented Reality, Taichung University of Education Library
[2] Li Yihong (2016). Virtual and augmented reality development trends and business analysis. Taipei City: Published by the Industry Research Institute of the Institute of Information Technology: Published by the Technology Division of the Ministry of Economic Affairs
[3] Zhu Huijun, 2013, augmented reality
[4] Chen, Hsiang & Wigand, Rolf & Nilan, Michael. (2000). Exploring Web users' optimal flow experiences. Information Technology & People. 13. 263-281. 10.1108/09593840010359473.
[5] Hsiang Chen (1999). Exploring Web users' optimal flow experiences. Computers in Human Behavior. Volume 15, Issue 5, 1 September 1999, Pages 585-608.
[6] Ronald T. Azuma.(1997).A Survey of Augmented Reality.Teleoperators and Virtual Environments 6, 4.355-385
[7] Action learning https://zh.wikipedia.org/wiki/%E8%A1%8C%E5%8B%95%E5%AD%B8%E7%BF%92
[8] Shenzhen Bailide Technology, 2016, Technology, URL:https://kknews.cc/tech/8mrxezq.html
[9] Wikipedia information processing theory https://translate.google.com/translate?hl=zh-TW&sl=en& u=https://en.wikipedia.org/wiki/Information_processing_theory&prev=search&pto=aue
[10] Wikipedia Double Code Theory https://en.wikipedia.org/wiki/Dual-coding_theory
[11] Learning Experience Pyramid http://www.vkmaheshwari.com/WP/?p=2332
[12] Yumeng Digital Technology Co., Ltd., 2019, is AR augmented reality advertising a future trend or a gimmick?, URL:Ghttps://www.arplanet.com.tw/trends/artrends/arad2019/
[13] Yumeng Digital Technology Co., Ltd., AR/VR Industry Development Trends in 2020, 2020 (Part I) AR Augmented Reality, URL:Ghttps://www.arplanet.com.tw/trends/artrends/2020artrend/

Smart Design, Science and Technology – Lam et al (eds)
© 2021 the Author(s), ISBN 978-1-032-01993-2

Research on the development strategy of community cultural prototypes: A case of a Xinpu persimmon dyeing workshop

Po-Lun Hou*
Department of Wood Science and Design, National Pingtung University of Science and Technology, Pingtung, Taiwan

Ming-Chyuan Ho
Graduate School of Design, Master & Doctoral Program, National Yunlin University of Science and Technology, Taiwan

ABSTRACT: The prototype of community culture is a concept and a method of development value, with operational elements. This study uses "Xinpu Persimmon Dyeing" as an example to explore this case through text analysis and field surveys, hoping to sort out a suggestion to guide community development. According to the conclusions that can be obtained in this study, the important development process and operational suggestions of "Xinpu Persimmon Dyeing" as a synonym for local characteristics have several points. 1. The starting point of the a cultural prototype with important local industries; 2. Early stage community participation and the establishment of community workshops; 3. The interaction and alliance between the workshop and other local industries; 4. Re-innovation of traditional skills and their own independent production and brand.

Keywords: Community crafts, Regional development strategies, Cultural prototypes, Crafts workshop.

1   INTRODUCTION

The failure and disappearance of Taiwan's traditional craftsmanship and the advancement of the times are closely related to the development of scientific and technological civilization. However, these declining industries are rich in locality and cultural affinity, showing ample national life and cultural connotations. In addition to the practical value of these traditional handicrafts, they can also cultivate the body and mind, as well as the value and function of social education. Among the industries that still exist today, the most prominent ones are direct industry associations, such as Yingge's ceramics and Sanyi's woodcarvings. This is also a proof of industry and a flourishing local handicraft industry. Xinpu Town, Hsinchu County, Taiwan has always used persimmon as its main cash crop. In recent years, one of the more distinctive industries in Hsinpu is the process of "persimmon dyeing." "Xinpu Persimmon" has become a synonym for Xinpu Town, just like Sanyi in Miaoli County is famous for wood carving, and "Sanyi Wood Carving" has become a synonym for local characteristics or brand impression. But unlike Xinpu persimmon dyeing, Sanyi woodcarvings have different origins since the Japanese rule. "Xinpu Persimmon Dye" became a synonym for Xinpu. It was only after 2010 that it gradually became one of the representative industries with local characteristics of Xinpu.

2   EXPERIMENTAL

In the development process of community characteristics or industry characteristics, how to set an operable "cultural prototype" is the most important task for local development. Only by incorporating cultural prototypes into community planning in the early stage has the advantage of sustainable development. In short, the development of local characteristic industries is not the same as the past development model. After qualitative changes, cultural creativity has become the core of local industries, and its development model is bound to change. Cohesion of consciousness is a very important factor in developing local characteristics.

Xinpu Persimmon is currently the most prominent and representative one of the emerging local characteristic industry development models. It is different from the local representative industry after the early foundry development model, or the local industry originally formed by persimmon dyeing. The most important key is their actions in the process of replacing persimmon cake with persimmon dye and shaping the cultural prototype of Xinpu persimmon dye that everyone can agree on. This case is currently the most representative example of Taiwan.

*Corresponding Author

DOI 10.1201/9781003188513-2

5

The case data studied is mainly primary data. The Xinpu Persimmon Dye Culture Association has been assisted by the National Taiwan Research and Development Center's Community Craft Support Program for many years. This case will refer to its plan application and achievement report to sort out its development context. Therefore, the first-level information is quite abundant. During the research process, it will be supplemented by field talks. The annual planners, Ms. Cai Linghui and Ms. Zhong Mengjuan, were interviewed.

The cultural industry is incorporated into the commercial consumer market through symbolization, and has become a special symbolic consumption model, and the consumer goods of this model also include ancient techniques such as traditional crafts. Japanese scholar Liu Zongyue (1941) pointed out that if the culture of craftsmanship did not prosper, all cultures would lose their foundation, because culture must first be living. In recent years, the overall policy has been focused on revitalizing the community, in which the concept of industrial revival has become the mainstream value concept. At the same time, it is also in line with Liu Zongyue's folk art and living handicrafts, including: "Regionalism-Folk Art Takes Roots in Everywhere," "Namelessness-Folk Art Made by Nameless Craftsmen," "Practicality-Folk Art Is the Starting Point," "Traditionality-Folk art inherits the technology and knowledge of ancestors," "Separated business-folk art is the product of several people working together," "Workability-Folk art is the result of meticulous hard work."

Local characteristic industries refer to townships, towns, and cities as regional units. Characteristic industries refer to economic activities developed based on local climate, geographical resources, historical stories, traditional skills and ethnic customs, and have unique, historical and cultural characteristics. This idea is derived from the Japanese One Village One Product (OVOP) campaign in Japan. This concept was proposed by the former governor of Japan's Oita prefecture, Dr. Morihiko Hiratsu, in 1979. That is, each township combines local characteristics to develop a distinctive craft or food specialty industry. The "local" category of local characteristic industries is mainly townships, towns, and cities. The developed products must have historical, cultural and unique characteristics. The contents promoted by the Small and Medium Enterprise Office of the Ministry of Economic Affairs of Taiwan are quite broad, including processed food, cultural crafts, creative life, local cuisine, recreational services and festival folk customs, such as Yingge ceramics, Hsinchu glass, Daxi dried beans, Yuchi black tea and other featured industries.

The "Related Laws on the Promotion of Traditional Crafts Industry" (referred to as "Transferring Law") promulgated in Japan in May 1974 requires that the following conditions must be met in order to be designated by the central government as national-level traditional crafts: 1. Mainly used for daily life; 2. The main part of the manufacturing process is manual; 3. Made with traditional technology; 4. The main raw materials produced are traditionally used raw materials; 5. Produced in a certain area and form the place of origin. This rule shows that the general outline of the traditional handicraft industry is similar to that of the local industry, so it is generally considered to be a part of the local industry, and it is called the "traditional local industry." Finally, the local industry also becomes an important catalyst for culture.

From the above literature, we can know that the initial formation of local characteristics came from the traditional skills, living needs, industrial development and the course of time and gradually formed a culture – the final gathering of people's connections to local emotions and experiences. This, in turn, creates cultural uniqueness. And such a niche must have the basic elements of a cultural archetype. There is a theory in social psychology, called the Mere Exposure Effect or the Repeated Exposure Effect which is a psychological phenomenon: people feel good because they are familiar with something. In social psychology, this effect is also known as the "familiarity principle." In addition to relying on "land, industry, people, culture and scenery" to establish local characteristic industries, another key factor is how to deeply embed the special impression of this place into the public's mind or subconscious. The so-called prototype is a natural tendency to experience things in a specific way. The prototype itself does not have its own form, but it behaves as we see it, the so-called "organizational principle." To establish a local characteristic industry based on this theory, we must create a sense of experience and imprint it in people's impressions. That is, to establish local characteristics, we must put forward our own concept of local characteristic prototypes. And through continuous repeated exposure to imprint the cultural symbol and form the characteristics of the place, which is Carl Gustav Jung's called "archetypes," the establishment of a collective consciousness. At the end of this study, it was defined as a "local cultural prototype" with local characteristics.

3 RESULTS AND DISCUSSION

This study will explore the development of Xinpu persimmon dyeing, and collect and study related activities and data of the sample in the past ten years. The main development strategy of Xinpu persimmon dyeing technology development will be looked at in each year. Participated in the community construction project of Hsinchu County in 2007, it continues to take persimmon dyeing culture as the main development strategy. In the study, the community followed the local "people, culture, land, production, and scenery" as the most community development indicators, and selected "persimmon dyeing" as the most preliminary key inspiration point for development. The first is skill cultivation, and the second is the creation of innovative values typical of local culture, that is, the development of experiences, stories, and creativity through multiple local related industry alliances as development strategies. The introduction of plant dyeing techniques in 2007, the combination of local cultural history and persimmon dyeing techniques in 2008, and the core cultural characteristics of Xinpu persimmon dyeing in 2009. The elements are based on the persimmon dyeing, which is calm and heavy, and has a personality like the Hakka people. Simple, stable, stiff and other characteristics, as a distinction from other plants. In 2010, the "Hsinchu County Persimmon Dye Culture Association" was officially launched. It established the strength of cultural connection for Hsinchu's Xinpu persimmon dyeing. It operates local culture-related businesses with NPOs and strengthens the use of local history and agricultural specialties and patterns. In 2011, Zhengcheng set up a professional workshop to lay an important foundation for the development of persimmon dyeing. The establishment of the workshop made the promotion of the association's culture smoother. After 2012, it has become a well-known local specialty industry, and has been cooperating with the local industry every year, providing persimmon dyeing technician training, providing on-site services, combining surrounding tourism resources, or integrating daily life as the theme, including the industrial. The reconstruction of the context of the square, the opening and extension of the concept of fashion and creativity, and the establishment of the aesthetics of life with Hakka persimmon dyeing made the local people agree more with the importance of persimmon dyeing culture in the Xinpu area. At the same time, measures such as bringing in tourists to promote consumption will be injected into the local economy. After 2016, we will open up diversified marketing channels: find stores for sale, manage established websites, and promote DIY kit tours for persimmon dyeing for schools, institutions, groups, travel agencies, etc., increase the visibility of Xinpu persimmon dyeing products and expand channel's set of points.

Synthesizing the development process of Xinpu persimmon dyeing, Xinpu persimmon dyeing has become a representative impression of local culture from the beginning of basic technical learning. The basic model of its cultural prototype was laid in 2009 to set the core cultural characteristics of Xinpu persimmon dyeing. Among them, the establishment of "Xinpu Persimmon Dyeing Workshop" is an important milestone in establishing its impression of Xinpu Persimmon Dyeing. The establishment of the professional workshop is more symbolic of the impression that Xinpu persimmon dyeing has taken root, and the various subsequent development strategies include combining the resources of the local ancestral temple culture with persimmon dye to drive the development of local culture and strengthen the cultural core of persimmon dye itself. Establishing good relationships and cooperative relationships with local stores have made the Xinpu Persimmon Dyeing House stronger. Under such mutual value, Xinpu Persimmon Dyeing House has launched its own brands and products. The channels have established a good relationship, and they

have brought out the integrity and order of Xinpu's development strategy.

4 CONCLUSION

The launch of Xinpu persimmon dyeing has a very strong cultural context with the local persimmon industry, and this cultural context comes from the special geographical environment of the local region. Xinpu Persimmon Dyeing Factory is a rising star, creating a new local characteristic industrial chain in the Xinpu área, further expanding the tourism benefits of the persimmon industry and strengthening the local cultural content. In this study, it is clearly understood that the relationship between the development model of the Xinpu Persimmon Dyeing Workshop and its development process is step-by-step, including the learning of skills, the establishment of cultural prototypes, the strengthening of cultural prototypes, the local connection of the elements and repeated exposure to the concept of Xinpu persimmon dyeing and its cultural archetype through this link, the link in this case being: "with the persimmon dyeing and calming as the element, its personality has the characteristics of simplicity, stability, stiffness, etc. Distinguish from other plants." It is important to make a clear definition and position to make the issue more cohesive. Among them, the establishment of the Hsinchu County Persimmon Dye Association and the Xinpu Persimmon Dyeing Workshop has a more direct connection between Hsinchu County, Hsinpu Town and Persimmon Dye. This effect makes it easier for the public to directly connect and imagine in the effect of repeated exposure. Therefore, it is suggested that the associations and workshops should be named after the local industries to better match the local characteristic industries. Among them, the establishment of a professional Xinpu Persimmon Dyeing Workshop is in response to the requirements of the counseling unit, but among them, the industrial awareness of workers is more concentrated. This is also a successful case of mutual cooperation between for-profit and non-profit units.

Summarizing the conclusions of this study, the impression of Xinpu persimmon dyeing can be fully imprinted in the minds and impressions of the public. In addition to having a complete prototype of local culture, it is to let people know the culture through repeated exposure. The operation strategy includes the following points:

A. The starting point of the cultural prototype with important local industries: the initial stage of Xinpu persimmon dyeing was the local persimmon industry, using its persimmon culture as the basic cultural prototype, redefining the local cultural prototype that belongs to persimmon dyeing, and closing it. It is in line with the spirit of the local Hakka and has obtained a high degree of spread and recognition.

B. Early stage community participation and the establishment of community workshops: because there

7

is a clear local cultural tone and because of the participation of enthusiastic people, the establishment of a community workshop has strengthened the operation base and foundation of the persimmon dyeing process, and also provides persimmon dyeing technician training, or provides resident service to promote local production. The participation of many people in the study helped the community to learn more about the culture of persimmons and to become an important base for the development of the community industry's external activities. It also seeks new economic development opportunities for local fruit farmers, so it has gained the recognition and participation of many local people, and it has laid the foundation for its success by building and strengthening the regional economy.

C. The interaction and alliance between the workshop and other local industries: The establishment of Xinpu Persimmon Dyeing Factory has played a role in strengthening the connection of local industries, including the reuse of ancestral temple culture, alliances with local shops, cooperation with restaurant operators, participation in various promotion activities, and promotion of local elementary and middle schools. The DIY experience has continuously exposed the impression of Xinpu persimmon dye in different industries. This is an important key strategy for promoting marketing. The purpose is to let everyone know about persimmon dye, not just marketing products. Through the persimmon dye, various industries are connected together, and the lifestyle industry is used to comprehensively let everyone realize the "persimmon dye culture," and share the benefits between them to form a close relationship industry. Showing extremely high strategic value.

D. Re-innovation of traditional skills and own independent production and brand: among the various traditional dyeing techniques, Xinpu Persimmon Dyeing Factory has become an important role-playing. It was established after Xinpu Persimmon Dyeing Cultural Association, but through the energy of the workshop to create a brand to develop life products to support the Hsinchu County Persimmon Dyeing Cultural Association. The event has become a strong

support system and development chip. The involvement of many designers has also provided diverse thinking in the community to develop different innovative techniques, so that traditional techniques can be re-innovated and indirectly affect regional innovation and enhance the aesthetics of local life.

Xinpu Persimmon Dyeing Workshop is a small and beautiful model of community craftsmanship. It plays a core role in the development of persimmon dyeing culture in Hsinchu County. Although it does not have the same high output value as large enterprises, it promotes regional cultural development through culture, including tourism. With living goods, Zongmiao culture and local DIY experience, it provides local employment opportunities and creates local value and indirect output value. This study only takes Xinpu persimmon dyeing workshop as a case study, and as a result has certain limitations, but it is also suggested that the development of community crafts or community industries should find its own local cultural prototype and shape its own local culture through the integration of local resources. The characteristics and positioning of the prototype allow the community to develop a clearer context and develop their own unique industries and characteristics.

REFERENCES

Jung, C. G., *The Archetypes and The Collective Unconscious,* Princeton, N.J.: Bollingen, 1934–1954, 2018.

Xingman, Huang, *Research on Local Industrial Development Measures in Japan,* Annual Economic Research Journal, Publisher: National Development Committee of the Executive Yuan, 263–288, 2004.

Yuanhong, Zhu, *"Cultural Industry: The Concept of Being Evaded Due to Prosperity"* Taiwan Industry Research, Issue 3, 11–45, 2000.

Zajonc, R.B. Mere A*Exposure: A Gateway to the Subliminal,* Current Directions in Psychological Science, 2011.

Zongyue, Liu, Xu Yiyi (Eds.): *craft culture,* Guangxi Normal University Press, 2006.

Traditional craft items inspection center diagnostic business report, Foundation Traditional Craft Industry Promotion Association, 2002.

Smart Design, Science and Technology – Lam et al (eds)
© 2021 the Author(s), ISBN 978-1-032-01993-2

Auxiliary design and research for elderly on setting up from sitting position

Cheng-Yu Wang*, Yung-Hsiang Tu & Ting-Shan Lai
Department of Industrial Design, Tatung University, Taipei, Taiwan

ABSTRACT: Social aging is an important global issue, and care for the elderly has become the focus of attention from all walks of life. With the increase of age, the articular cartilage has obvious changes in structure, molecules, cells and mechanical mechanics, which increases the vulnerability of tissue degradation. In addition to the inconvenience of walking in life, the elderly have difficulty in sitting up which is a big problem. At present, there are not many mobile assistive devices with stand-up assistance on the market, and the evolution has been slow. This study asked 10 users with an average age of 90 years old, including 2 males and 8 females, to ask questions about the use of assistive devices through questionnaires to understand the needs of users. The question is mainly divided into 3 types of assistive devices, measured by 5 points Likert, to understand the user's opinions on technology acceptance (easy to use, useful, attitude, willingness to act), the results show that design A To reduce the burden, weight affects the use of these two needs to be improved. Design B is inferior in the simplicity of folding, but better in portability. Design C's attributes of assisting action, increasing willingness to use the product, feeling easy to use, being worth using, helping to do things, etc., all need to be improved. In the future, we will provide better design solutions for the elderly through participatory design methods at several meetings. Then, arrange comparative experiments to test the practicability of the new design.

Keywords: Elderly, Auxiliary design, Stand-up assistance, TAM.

1 INTRODUCTION

Population aging is an important global issue. In today's world, most people are expected to live to be over 60 years old, which is unprecedented (World Economic and Social Survey 2007 2007). By the end of the 2030 Healthy Aging Decade, the number of people aged 60 and over will increase by 56%, from 962 million (2017) to 1.4 billion (2030). By 2050, the global elderly population will more than double to 2.1 billion (WHO 2019). The decrease in the mortality rate of the elderly is the main reason for the increasing average life expectancy (Christensen et al. 2009). Most health problems of the elderly are caused by chronic diseases, many of which can be prevented or delayed by adopting healthy behaviors. In fact, even at an advanced age, physical activity and good nutrition are of great benefit to the health and well-being of individuals. Other health problems, especially when detected early, can also be effectively controlled. Even for the elderly with declining ability, a good supporting environment can ensure that they can still go where they need to go and do what they need to do (Global Report on Aging and Health 2016). Care for the elderly has become the focus of attention from all walks of life. With the increase of age, the articular cartilage has obvious

changes in structure, molecules, cells and mechanical mechanics, which increases the vulnerability of tissue degradation (WHO 2019). Muscle quality often begins to decline with age as it peaks in early adulthood, and this decline can lead to decreased strength and musculoskeletal function (Cruz- Jentoft et al. 2010). Aging causes the body to deteriorate. It is difficult to use drugs and surgery to improve their situation, so the elderly need to rely on mobility aids to assist with everyday life (Huang 2005).

Walking is the most common movement of a daily activity (Shurr & Cook 1990), but for the elderly, getting up and walking look like easy operations, but for them these activities can be very difficult. Many elderly people may experience accidents such as falls when they are walking due to poor physical function (Goodpaster et al. 2006). There are 2.99% of those 55 to 64 years old who cannot sit on a chair without standing with their hands, and this figure rises to 16.41% when they are over 65 years old (Ministry of Health and Welfare 2019). Recent studies have found that some scholars have entered into the discussion about the changes of aging and onset and sitting down. Many studies have found that it takes a lot of time for elderly people to start and sit down. This is due to the change in the torso angle during the transition from the starting position to the sitting position, which increases the risk of falling (Lehtola et al. 2006; Leung & Chang 2009).

---

*Corresponding Author

DOI 10.1201/9781003188513-3

Auxiliary appliances are any external appliances (including devices, equipment, instruments or software) that are specially produced or generally available, whose main purpose is to maintain or improve an individual's physical function and independence and thereby promote the individual's well-being (List of assistive devices 2016). At present, the stand-up assisting tools on the market can be roughly divided into fixed type, mobile type and other assisted types. However, there are not many styles of mobile assistive devices with stand-up assistance.

This study will use the questionnaire method to let seniors fill out the questionnaire to understand the seniors' ideas and needs for existing stand-up assistive devices.

2 INVESTIGATION

Three kinds of aids have taken up the market function in receiving the survey sample, as is shown in Table 1.

The purpose of the questionnaire is to understand the views of the elderly on stand-up assistive devices. The content of the questionnaire is based on the Technology Acceptance Model (TAM). The topics are divided into 4 major aspects: "feel easy to use", "feel useful," "Attitude," and "Willingness to act" over a total of 15 questions. Likert's 5-point scale (divided into "strongly disagree," "disagree," "normal," "agree," "strongly agree," measured by 1–5 points) was used for measurement. The purpose is to analyze the problems with assistive devices and to provide a theoretical basis for the subsequent experiments.

The questionnaire design is divided into 3 categories:

A. The first category is basic information, including: gender, age and type of residence.
B. The second category is the acceptance question:
(1) It feels easy to use: 1. The walker is easy to operate, 2. The walker is easy to fold and close,

3. The walker is easy to get started, 4. The walker is easy to carry.
(2) Feeling useful: 1. Walkers can let me go further, 2. Walkers reduce my burden, 3. Walkers help me accomplish what I want to do.
(3) Attitude: 1. I am willing to use the walker, 2. I think the walker is easy to use, 3. I think the walker is worth using, 4. I am happy to use the walker.
(4) Behavioral intentions: 1. Whether the function affects my willingness to use the walker, 2. Whether the weight affects my willingness to use the walker, 3. Whether the aesthetics affects my willingness to use the walker, 4. Whether the size affects my willingness to use Walker.
C. The third category is open question
Do you have any suggestions about the design of the walker?

After collecting the data, we performed a single T test in the T test. The higher score of 4 was divided into test values. Then we checked whether the average score of all question items is significantly different from 4 points.

3 RESULTS

A total of 10 senior citizens were interviewed in this questionnaire, including 8 males and 2 females. Age distribution in the 85 year to 95 year range was: 2 for 85 years old, 1 for 89 years old, 1 for 90 years old, 3 for 91 years old, 2 for 92 years old and 1 for 95 years old. All people surveyed live in a nursing facility.

According to the results in Table 2, we can see:

(1) Feels easy to use

In terms of "simple folding," the agreement of design A has no significant difference from 4 points, but plan B is significantly lower than 4 points (B:$t = -2.753$ $p = 0.22$). The satisfaction of the B design scheme on the "easy to carry" project was significantly higher than 4 points (B:$t = 2.449$ $p = 0.037$).

(2) Feeling useful

In terms of "assisted action," The agreement degree of the C design plan should be significantly lower than 4 points (C:$t = -2.449$, $p = 0.037$). In terms of "reducing the burden," the design agreement degree of A is significantly lower than 4 points (A:$t = -3.857$, $p = 0.04$) In "help to do things," the agreement of the C design plan was significantly lower than 4 points (C:$t = -2.714$, $p = 0.024$)

(3) Attitude

In "willing to use," "feels good" and "worthwhile to use" on these three areas, C design of the consent of all significant with less than 4 points (respectively $t = -2.806$, the $p = 0.021$; $t = -2.689$, $p = 0.025$ and $t = -2.753$, $p = 0.022$). There is no obvious difference between the two design schemes, A and B, in the attitude sub-item and 4 points.

Table 1. Design of questionnaire survey.

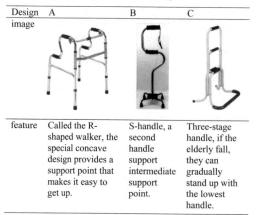

Design	A	B	C
image			
feature	Called the R-shaped walker, the special concave design provides a support point that makes it easy to get up.	S-handle, a second handle support intermediate support point.	Three-stage handle, if the elderly fall, they can gradually stand up with the lowest handle.

Table 2. Survey results of three types of stand-up aids.

| | Single T test (test value $= 4$) $\alpha = 0.05$ | | |
Aids	$X < 4$	$X = 4$	$X > 4$
A	Reduce the burden $p = 0.004$ The weight effect uses $p = 0.045$	Easy to operate Easy to learn	
B	Simple folding $p = 0.022$	Pleasant to use	Easy to carry $p = 0.037$
C	Help action $p = 0.037$ Willing to use $p = 0.021$ I feel easy to use $p = 0.025$ It is worth using $p = 0.022$ Help doing things $p = 0.024$	Function affects use Appearance affects use Size affects use	

Note: X represents the average number of variables

(4) Willingness to act

In terms of "weight affects use," design A's consent level is significantly lower than 4 points ($t = -2.333$, $p = 0.045$), and there is no significant difference between the agreement degrees of other items and 4 points.

Summarizing Table 2, Design A needs to be improved in order to reduce the burden and affect the use of weight. Design B is inferior in the simplicity of folding, but better in portability. Design C, in the processes of helping action, willing to use, feel easy to use, worth using, helping to do things, etc., needs to be improved.

In the free answer "what are your suggestions about the above-mentioned assistive devices?" three people answered, and two of them said that the assistive devices were not convenient when assisting them to go upstairs. Another said that it is an indispensable tool that helps him use the toilet at night.

4 CONCLUSION

In the process of testing, it was found that when the elderly faced unfamiliar things, they often showed a resistant attitude. Especially when the new things and the things you are familiar with cannot be referred to, they will say "I haven't used them, so I don't know." They need to be guided slowly, such as with the C design, they can be asked: "do you think this is like a handrail, do you think this handrail is easy to use?" So that the elderly can answer. Some elderly people have a mentality of resistance to assistive devices and are reluctant to use them. Although admitting that assistive devices do help them to walk, once rehabilitating when they can walk, they are reluctant to use them. Some seniors feel that assistive devices are for elderly people with mental retardation.

The A design is larger than the B and C designs, and it feels heavier (even though its weight is usually lower than that of the crutches) The A design is mainly aimed at elderly people who have difficulty walking. When the elderly can use crutches, they tend to use crutches.

The B design is a revised version of the crutches, so the elderly can understand its purpose. The B design has its advantages in terms of portability. Because it has no design considerations for collapsing, it is not ideal for simple folding.

C design is the concept of adding a handrail on the basis of crutches, mainly used at home. Because its main function is to stand up, so one of the evaluation items is added. It is because it is used at home so that the elderly are rarely seen. It is an assistive device that the elderly are not familiar with. As I wrote earlier, the old man shows resistance to unfamiliar things. So the score for many terms is not high, and he is not willing to use it.

Through this survey, we learned the elderly's thoughts on stand-up assistive devices, understood the opinions of the elderly on the existing assistive devices on the market, and have hinted at the future designs of stand-up assistive devices. Because C design has the most improvement points, there is a lot of room for improvement. The design direction is expected to be based on C design and combined with the advantages of the other two designs.

REFERENCES

Christensen K, Doblhammer G, Rau R, Vaupel JW. Ageing populations: the challenges ahead. Lancet. 2009 Oct 3; 374(9696): 1196–208. doi:http://dx.doi.org/10.1016/S0140-6736 (09) 61460-4PMID: 19801098

Cruz-Jentoft AJ, Baeyens, JP, Bauer JM, Boirie the Y, T Cederholm, Landi F., Et Al.; European Working Group ON Sarcopenia in Older People Sarcopenia:. European consensus ON Definition and Diagnosis: The European Working Group of the Report ON Sarcopenia in Older People. Age Ageing. 2010 Jul; 39 (4): 412–23. doi: http://dx.doi.org/10.1093/ageing/afq034 PMID: 20392703

Goodpaster, BH, Park, SW, Harris, TB, Kritchevsky, SB, Nevitt, M., Schwartz, AV, Simonsick, EM, Tylavsky, FA, Visser, M., & Newman, AB (2006). The Loss of

Huang, YC. (2005). A Study on an Assistant System for Elder. Bimonthly Bulletin of Recreation and Mobility Industries, 20, 1–11.

Lehtola, S., Koistinen, P., & Luukinen, H. (2006). Falls and Injurious Falls Late in Home-Dwelling Life. Archives of Gerontology and Geriatrics, 42, 217–224.

Leung, CY, & Chang, CS (2009). Strategies for Posture Transfer Adopted by Elders During Sit-to-Stand and Stand-to-Sit. Perceptual and Motor Skills, 109, 695–706.

Ministry of Health and Welfare (2019), Survey Report on the Status of the Elderly in 106 in the Republic of China, p29

Shurr, DG, & Cook, TM (1990). Prosthetic and Orthotics, Appleton and Lange, New York.

Skeletal Muscle Strength, Mass, and Quality in Older Adults: The Health, Aging and Body Composition Study. The Journals of Gerontology Series A: Biological Sciencesand Medical Sciences, 61, 1059–1064.

WHO, *Global Report on Aging and Health*, https://apps. who.int/iris/bitstream/handle/10665/186463/9789245565 048_chi.pdf;jsessionid=53A9A0976411CE002EED6061 3F52EAEE? Sequence = 9 P4, 2006

WHO, *WHO list of key assistive devices*, https://apps.who. int/iris/bitstream/handle/10665/207694/WHO_EMP_PH I_2016.01_chi.pdf? sequence = 3, P1, 2016

WHO, *2020–2030 Healthy Aging Decade for Action*, https:// www.who.int/docs/default-source/documents/decade-of-health-ageing/decade-ageing-proposal-en.pdf?sfvrsn = b0a7b5b1_12, P1, 2019

Smart Design, Science and Technology – Lam et al (eds)
© 2021 the Author(s), ISBN 978-1-032-01993-2

Research on laser processing effect of Mother-of-Pearl Inlays after softening

Po-Lun Hou*, Shiao-Yu Chang & Yu-Tin Su
Department of Wood Science and Design, National Pingtung University of Science and Technology, Pingtung, Taiwan

ABSTRACT: The Mother-of-Pearl Inlays process is a highly cultural and artistic creation, which has a wide range of applications. It can be seen in all lacquer arts and furniture, and it is applied with Mother-of-Pearl The price of furniture with Inlays technique is much easier than that of ordinary furniture, which shows that its value is extraordinary. However, the selling price of Mother-of-Pearl Inlays furniture has also limited its development. Especially, Mother-of-Pearl Inlays technology often relies on highly labor-intensive operations, especially in Mother-of-Pearl Inlays pattern cutting is highly technical. In this study, through traditional processing methods, we tried to use Mother-of-Pearl Inlays before laser cutting as the subject of discussion, and developed into modern processing techniques to reduce dependence on manpower, thereby reducing its cost Price. Laser cutting under normal conditions will damage the flatness and embrittlement of the edge of the Mother-of-Pearl Inlays. This study hopes to make Mother-of-Pearl Inlays easier through natural softening Laser cut. Finally, it was concluded that Mother-of-Pearl Inlays soaked in the natural white radish purce has the best effect, and it is the least energy-consuming and consistent with the sustainable spirit of agricultural waste recycling.

Keywords: Mother-of-Pearl Inlays craft, Mother-of-Pearl Inlays softening, laser engraving, craft creation

1 INTRODUCTION

The Mother-of-Pearl Inlays(螺鈿) process has been developed for more than 3000 years. The main furniture style of traditional Mother-of-Pearl Inlays furniture is Chinese style furniture. Mother-of-Pearl Inlays is commonly known as "shell Inlays" in China. It is a decorative process that encloses shells-like flakes according to a preset pattern and inlaid on the carving. According to data, the earliest works of Mother-of-Pearl Inlays originated from the Western Zhou Dynasty. During the Tang Dynasty, Mother-of-Pearl Inlays was introduced from China to North Korea and Japan, affecting the development of lacquer crafts in East Asia. Around the 14th century, it spread to Ryukyu and Southeast Asia with sea trade. Furniture Inlaysing process refers to the process of Inlaysing various texture materials on furniture to form various patterns. The origin of the furniture Inlays technology is relatively early, the technology has taken shape in the Tang Dynasty, and it was more widely used in the Song Dynasty. The materials used for the mosaic pattern are mostly mother-of-Pearl Inlays, gold and silver, porcelain, and marble. The objects made are screens, tables, chairs, and boxes. After the restoration of Taiwan, Mother-of-Pearl Inlays furniture was produced in Hsinchu, Zhongli, and Taoyuan, supplying domestic and foreign sales. The change of the market and the difficulty of obtaining materials have now made it quiet and declining. Traditional mother-of-pearl Inlays are made by hand cutting, which is time-consuming and labor-intensive. The basic wage level of the modern labor force is gradually increasing, and traditional production will make the cost very high. This research hopes that the high labor cost of processing Mother-of-Pearl Inlays can be reduced by the aid of modern technology, thereby reducing the manufacturing cost of Mother-of-Pearl Inlays Furniture.

2 EXPERIMENTAL

The production materials produced by Mother-of-Pearl Inlays are mainly shells. There are many types of shells, the colors are bright and gorgeous, and there is pearly luster; the crystal bright surface material in the shells uses the noble texture, after grinding and cutting The resulting shell flakes are used as mosaic objects. It can be roughly divided into Thick Mother-of-Pearl Inlays and Thin Mother-of-Pearl Inlays: Thick mother-of-pearl Inlays are usually white or ivory, also call as "Hard Shell Fill(硬螺填)". Thin Mother-of-Pearl Inlays can show red, pink, blue and other beautiful luster, colorful and gorgeous, also known as "Soft Shell Fill(軟螺填)". The decorative technique of making patterns such as figures, birds, beasts, flowers and plants from shell shells and inlaid on the surface of carved or lacquered utensils is called Mother-of-Pearl Inlays. The so-called Mother-of-Pearl Inlays are sculpted by the shell clam, and carved into the geometric figures

*Corresponding Author

DOI 10.1201/9781003188513-4

that the craftsman wants to express, and then inlaid on the object. Chinese word "鈿" means "Floral Twinkle". It is a precious ornament on the golden flower, decorated with brilliance like jade and spots like flowers. This process originated very early and has become popular in the Zhou Dynasty. From the physical view of the existing Tang Dynasty's Mother-of-Pearl Inlays, it was already of a very high level. Because China divides them into thick Mother-of-Pearl Inlays and thin Mother-of-Pearl Inlays. The shell is inlaid on a smooth lacquer surface, which is particularly eye-catching.

Skill sequence of Mother-of-Pearl Inlays:

1. Tracing manuscript: According to the needs of the artwork, select the Mother-of-Pearl Inlays corresponding to the hue, and dip the ink to trace the outline of the artwork on the septum.
2. Cutting: Use a pen or a needle knife to cut and shape the Mother-of-Pearl Inlays. Carve a pattern on the Mother-of-Pearl Inlays, and pay attention to prevent the diaphragm from breaking.
3. Paste and flatten: Useing glue as an adhesive, paste the Mother-of-Pearl on the smooth and medium-painted surface, and use a small iron (temperature controlled at 60-70 degrees, to prevent Mother-of-Pearl scorch) to make the cymbal flat against the carcass.
4. Wipe: Dip hot water with a pen to scrub excess glue on the surface.
5. Fixation: Rub the Mother-of-Pearl Inlays gap thinly with raw lacquer or black lacquer to fix it.
6. Lacquer: After leaving for two days, allow the surface to fully dry, and then paint the whole body several times.
7. Grinding: first scrape the paint film on the cymbal with the tip of a bamboo slip, and then grind the Mother-of-Pearl Inlays pattern to make the texture smooth.
8. Wipe paint and polishing.

The mother-of-Pearl Inlays technique can be divided into Thick Mother-of-Pearl Inlays technique and Thin Mother-of-Pearl Inlays technique:

1. Thick Mother-of-Pearl Inlays technique: Pattern drawing, cutting, designing patterns and sticking to the shells → bow saw cutting to remove the graphics → furniture and products to be inlaid part inlaid → sanding the shells and trimming the contours to make inlays → engraving. The operation methods are further divided into: flat inlay, Embossed inlay, Embossed paste and other methods. The main tools required include saw table, wire saw (hand saw), file, carving knife and so on.

   A. Flat inlay: Mother-of-Pearl Inlays are sawn and trimmed into the body, and the surface is flush with the body.
   B. Embossed inlay: After inserting into the body, the surface of the mother-of-Pearl Inlays is higher than the surface of the body.
   C. Embossed paste: No Inlay behavior, stick directly to the body.

2. Thin Mother-of-Pearl Inlays technique: Pattern drawing → Draw the pattern on Mother-of-Pearl → remove the pattern → paste on the lacquer carcass → the carcass part is completed. Commonly used tools are cutting pads, special needle hammers, utility knives, carving knives, tweezers, bamboo sticks, small irons, paperweights, etc.

Throughout the entire Mother-of-Pearl Inlays process, cutting a Mother-of-Pearl is extremely easy to fail, because Mother-of-Pearl itself is quite fragile, and a little carelessness will make the whole piece The mother-of-Pearl is broken and unusable, the cost of processing technology is high, and it is time-consuming and time-consuming, resulting in very high labor costs. Mother-of-Pearl is thin and brittle. If it is cut directly, the edges are prone to irregular edge damage. Therefore, it must be softened before cutting. The softened scallop is easier to cut smooth curves and is not easy to crack. The thin shell is about 0.25mm. Before use, it must be treated before it can be used as a snail of Mother-of-Pearl Inlays. In the traditional method of use, the mother-of-Pearl pieces are soaked in radish water before cutting to soften the shells and less likely to break during cutting.

Therefore, there are two types of cutting methods in the traditional Mother-of-Pearl:

1. Corrosion method: Brush diluent (Mother-of-Pearl Inlays) tablets with dilute hydrochloric acid. The residual hydrochloric acid needs to be washed with water.
2. Cutting method: Use a needle knife to cut flat, large curved patterns, such as square rectangles, etc. After softening the sepals, you can use scissors to cut the shapes, but the small curved portions of the patterns are not applicable.

The corrosion method is not environmentally friendly, and it is extremely likely to cause occupational injury in the process. The cutting rule is time consuming, the processing technology is high, and the labor cost is higher.

In this study, we will focus on the processing using the Mother-of-Pearl Inlays technique and discuss the best parameters for the softening of Mother-of-Pearl Inlays. The corrosion method is not environmentally friendly, and it is extremely likely to cause occupational injury in the process. The cutting rule is time consuming, the processing technology is high, and the labor cost is higher.

In this study, the feasibility of engraving the processing method of Mother-of-Pearl Inlays through a laser engraving machine will be explored, and other methods of softening the environment friendly Mother-of-Pearl Inlays will be discussed. In the experiment, the comparison method between the experimental group and the control group was used to explore the best parameters of laser processing for different softening materials and the softening time of the mother-of-pearl. Laser cutting uses the same power and speed to process Mother-of-Pearl flakes made from abalone shells to explore the effect of edge coking. In the processing of Mother-of-Pearl Inlays, the cutting of the edges is very important and will affect the cost quality. Therefore, obtaining a good edge is a very important process. Therefore, the important factors discussed in this study include:

1. Mother-of-Pearl Inlays softening time is the best parameter for edge coking control during laser cutting.
2. Comparison of the softening effect of Normal water, Hot water, Radish puree softening material on Mother-of-Pearl.
3. Laser replaces people's proposal for Mother-of-Pearl.

Experimental process:

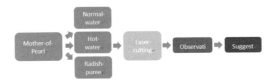

Figure 1. Experimental process.

The softening experiment method will use the tradional and normal Water, Hot water and Radish puree instructions:

1. Immerse directly in water at normal temperature, and then immerse Mother-of-Pearl in it. Take samples at 15 minutes and 30 minutes, respectively, and use a laser engraving machine to perform cutting tests.
2. Immerse in the continuously heated hot water, and then immerse Mother-of-Pearl in it. Take samples for 15 minutes and 30 minutes respectively, and then use a laser engraving machine to perform cutting tests.
3. The radish was peeled and prepared into a mud-thin shape, and Mother-of-Pearl was immersed in it. Samples were taken at 15 minutes and 30 minutes, respectively, and a laser engraving machine was used for cutting tests.

3 RESULTS AND DISCUSSION

Under the condition of the control group, the cutting effect without immersion is used as an evaluation benchmark. Although cutting can be performed, the

Table 1. Table of softening effect picture of abalone shell.

Method	Control group	Normal water		Hot water (100℃)		White Radish Puree	
Laser	Cut						
Laser Power/Speed	10W/350S						
Soaking time(minutes)	0	15	30	15	30	15	30
Picture							

Table 2. Record of the effect of abalone shell softening.

	Method	Soaking time (minutes)	Result
Control group	Not soaked	0	1. Scorch edges more 2. Cutting edges are not slippery. 3. Corners and edges are brittle.
Experimental group	Normal water	15	1. scorch edges more. 2. Cutting edges are not slippery. 3. Corners and edges are brittle
		30	1. scorch edges normal. 2. Cutting edges are not slippery 3. The corners are brittle and the edges are less brittle.
	Hot water (100°)	15	1. scorch edges more 2. Cutting edges are not slippery. 3. Corners and edges are brittle.
		30	1. scorch edges very few 2. Neat edges 3. The corners are slightly brittle and the edges are not brittle.
	White Radish Puree	15	1. scorch edges less 2. Cutting edges are not slippery. 3. Corners and edges are brittle.
		30	1. scorch edges very few. 2. Neat edges. 3. The corners are slightly brittle and the edges are not brittle.

cutting effect is not ideal. According to the above literature, it is known that the evaluation conditions should meet 1. The edges should not be coked, and Mother-of-Pearl's original color, 2. The cut edges must be neat, and then they can have good fit when they are fitted to the carcass. 3. Mother-of-Pearl cannot be embrittled at the edges and corners of the cut. After embrittlement, a larger tolerance will be generated for the fitted edge, resulting in a reduction in quality. In the experiment, these three points are used as the evaluation points to observe the situation after laser cutting.

The experimental results show that Mother-of-Pearl Inlays need to be softened to achieve a good cutting effect. Taking abalone shells as an example, hot water and white radish puree are the best. The effect is sorted as follows: soaking radish puree for 30 minutes = slow boiling in hot water for 30 minutes > hot water for 15 minutes > soaking radish puree for 15 minutes. In the experimental results, the effect of white Radish Puree soaking for 30 minutes and slow boiling in hot water for 30 minutes is quite acceptable, especially after 30 minutes of White Radish Puree soaking. By laser cutting, the edges have very little coking. If you want to get better results, you can fine-tune the laser power to find better results. The flatness on the edges is the best in each group. In the corner, although the embrittlement is small, it is already the better effect in each group, so it still has its reference value. Therefore, if laser processing is used in the design of the Mother-of-Pearl Inlays, the contour lines should be designed to avoid the smallest angle design, and the arc should be used to draw smooth paths. Achieve good cutting results. The softening of snail shells under the condition of maintaining hot water at 100°C is more severe than the other two groups, and it consumes more energy. It will also increase the cost of production, and more evaluation conditions are bound to be made.

4 CONCLUSION

After comparing this experiment with the control group, it is verified that a good edge trimming effect will affect the product cost and quality. It is suggested that Mother-of-Pearl needs to be softened before laser processing to achieve better results and improve the integrity of the finished product. According to the results, the hot water heated in the pool and the white radish puree were the best for softening the Mother-of-Pearl Inlays. Considering the convenience of processing in the factory, the use of the hot water method is a good choice. In addition to being convenient and easy to use, it is also good for recycling. However, what is derived from this method is the need for electric heating and continuous heating, which is relatively energy consuming and not environmentally friendly. From the perspective of environmental

protection, white radish can be collected as softening raw material for white radish used in markets or restaurants. In addition to the effect of not wasting, it also achieves the purpose of reuse and saves heating energy. Therefore, if you want to pursue integrity and recycling and reduce costs, it is recommended to use the Radish Puree method, in addition to manufacturing Mother-of-Pearl Inlays furniture, you can also contribute to the environment. However, the White Radish Puree law needs to pay attention to the hygienic and oxidized odor problems of recycling. In addition, because the thickness of the Mother-of-Pearl sheet is only about 0.25mm, it is recommended to use double-sided tape or paper tape to fix it to acrylic or wood with a small thickness when laser cutting to prevent the finished product from falling directly to the machine. As for the optimal parameter setting of Laser processing for Mother-of-Pearl edge coking control, because the setting parameters of each laser engraving machine are different, it is impossible to make reference for the optimized values here. But the thin Mother-of-Pearl Inlays film itself is very thin. Even if the processing machine is different, it is sufficient to set the processing to minimize the power parameters according to the characteristics of the machine. In the whole result, laser processing replaces manual cutting of Mother-of-Pearl is very feasible. Technology-assisted cutting can effectively cut Mother-of-Pearl instead of manual cutting, which can more effectively reduce the failure rate of Mother-of-Pearl cutting. And increase production efficiency, can save more labor costs. The softening effect of white radish puree can also make the whole process smoother and meet the requirements for mass production. Although it is not known what the opportunity for white Radish Puree to soften the shell, for the field of biomass utilization, Radish does take into account environmental protection and the goal of promoting sustainable development, which is a new development in the future. One of the biomass materials is that we can continue to explore research topics in the future.

REFERENCES

Ho Rongliang, 2005. *Lai Gaoshan Lacquer Creation Research Album*. National Taiwan Crafts Research Institute, Caotun Town, Nantou County

Lai Zuoming, 2005. *Taiwan's lacquer art in the endless world*. Art &Collection Group, Taipei.

Lin Meichen, Hong Wenzhen, Wang Xianmin, 2001. *Wang Qingshuang Lacquer Creation 80 Review*. National Taiwan Crafts Research Institute, Caotun Town, Nantou County.

Weng Xude, Huang Lishu, Jian Rongcong, 2001. *Taiwan lac querware cultural heritage-Penglai lacquerware*. Nantou County Folk Cultural Relics Society, Caotun Town, Nantou County.

Smart Design, Science and Technology – Lam et al (eds)
© 2021 the Author(s), ISBN 978-1-032-01993-2

VR interface design for promoting exercise among the elderly

Yu-Min Fang* & Yen-Jung Huang
Department of Industrial Design, National United University, Miao-Li, Taiwan

ABSTRACT: By the 21century, the elderly population will exceed the total population of the last century. The issue of caring for the elderly has resulted in the need for health care, assisted living, and independence of the elderly. Previous research has confirmed that introducing health technology at an early stage can encourage the elderly to continue their activities, promote their independence and reduce the cost of social care. In this study, which applies the advantages of virtual reality (VR) technology, Unity 3D software is used to design VR dumbbell motion software for the elderly This software can provide positive excitation and enable the elderly to maintain continuous motion. The objectives of this research are as follows: (1) to construct VR prototypes; (2) to test their usability (i.e., operating efficiency, visibility, and error rate); and (3) to study the emotional response of interfaces. This study adopted the Self-Assessment Manikin to measure emotion and the questionnaires for user interaction satisfaction to measure usability Interface experts were also invited to conduct experiments, fill in questionnaires and provide interviews. The comparison of the VR application programs with physical dumbbells yields the following conclusions. The VR application programs are more pleasant. The real water dumbbell interface feels slightly stronger. The interface learning performance of the VR application programs is better. The preliminary conclusions drawn from this study can be used as a reference for designers Scholars are expected to conduct further explorations.

Keywords: The Elderly, Virtual Reality, Social Interaction, Usability, Emotional Design, Exercise for Older Adults

1   INTRODUCTION

The United Nations forecast that the number of elderly people in the 21st century will exceed the total population of the last century [1]. The physical and mental functions of the elderly gradually decline with an increase in age, and thus, special product design principles should be applied to the elderly [2–4]. Moreover, the issue of caring for the elderly has resulted in the need for health care, assisted living, and independence of the elderly [5]. The following questions are worth exploring How can elderly people who are suffering from emotional loneliness be assisted? How can appropriate forward-looking interactive technology from various existing devices be provided in accordance with the needs of integration? How can the elderly be motivated to perform appropriate exercises? How can a software interface design be provided? In what ways can the independence of the elderly be promoted?

Previous research has confirmed that introducing health technology at an early stage can encourage the elderly to continue their activities. Doing so will improve their quality of life and delay the degree of functional decline and disability caused by aging [6], further promoting their independence and reducing the cost of social care [7].

Virtual reality (VR), which is alternately known as virtual environment, is described as an environment simulated by a computer [8–10]. It is a 3D virtual world that can be accessed through computer simulation, providing users with multiple sensory stimulations. Users feel as though they are actually experiencing the scene they are watching and can observe objects in 3D space instantly and indefinitely. Given the advantages of VR technology, positive stimulation can be provided to keep the elderly in constant motion and delay aging. The objectives of the current research are as follows. First, VR prototypes are constructed, and the design principle of an active excitation interface for the elderly is proposed through experiments. Second, the background and experience of the user group are compared, and the usability of the prototypes (i.e., operating efficiency, visibility, and error rate) is tested. Third, the emotional responses of the interfaces are studied, including emotional valence and arousal.

This study adopted the Self-Assessment Manikin (SAM) to measure emotion. SAM is a visual scale used to measure the emotional dimension; it was proposed by Mehrabian and Russell [11]. SAM is a semantic scale and includes three emotional aspects: emotional valence, arousal, and dominance SAM measures 18 different emotional states and uses emojis as a test to avoid cognitive differences caused by language barriers. After long-term and extensive testing, this scale is found suitable for measuring emotions using an

---

*Corresponding Author

DOI 10.1201/9781003188513-5

interactive computer interface [12,13]. Only the nine-point scales of two emotional aspects (i.e., emotional valence and arousal) were adopted in the current study. Dominance is irrelevant to this study; hence, it is eliminated. In the final questionnaire used in this study, opposite semantic adjectives were added to the left and right ends to make interpreting the answers easier for the subjects.

To measure usability, this study adopted the questionnaire for user interaction satisfaction (QUIS) QUIS measures the subjective satisfaction of a system to the user's human–machine interface, including measurements of visibility, system information, learning, and system functionality [14–16]. QUIS can be modified in accordance with the study samples, each with a title of a sevenorder Likert scale.

For its experimental equipment, this study adopted the Unity 3D (Unity Technologies) software to present interaction VR. Market analysts expect the economic impact of VR and augmented reality (AR) to increase from 58 billion (low adoption rate) to 205 billion (high adoption rate) in 2020. The market size of VR software in 2018 was estimated at 48 billion USD. At present, VR platforms include HTC Vive, Oculus Rift, PSVR, Samsung Gear VR, Google Daydream, and Microsoft HoloLens. Examples of development software with a high market share are Unity 3D, Unreal, and CryEngine. VR and AR experiences require different tools and development environments. For example, when Unity 3D is used to develop VR games, a fixed brand headset display (e.g., HTC Vive, Oculus Rift, or PSVR) should be specified. If an AR application is required to be developed, then a cross-platform toolset (e.g., ReactVR) can be used. Otherwise, an iOS or Android device should be further located. If the development of a VR experience is implemented for HoloLens, then it can start with Unity 3D, and then the application program can be tested and deployed using Visual Studio.

Considering commonality and universality, Unity 3D is selected in this study. This software is equipped with a professional 3D game engine that exhibits the characteristics of a cross-platform, efficient optimization, and game screen dyeing effect. Unity 3D currently has a wide range of applications ranging from mobile games to large-scale online games, from serious games to e-commerce, and then VR. Games developed using this software can achieve rapid real-time execution speed, has good hardware equipment, and can calculate more than millions of polygons per second. The built-in particle system can control particle color, size and movement trajectory; it can create rain, flame, dust, explosion, fireworks and other effects [17].

2 EXPERIMENTAL

In this experiment, dumbbells were selected as the subject of detection The VR application was developed using Unity 3D and was compared with the water

dumbbells of the entity. Measurements included ease of use (interface design) and emotional arousal and titer scales. Four participants were invited to take part in this study.

2.1 Sample design

Related literature was discussed, and experimental samples were made by referring to existing VR sports games. The samples of the computer version of the motion VR software were tested using the headset display device HTC Vive and the self-developed software Unity. The tested samples were compared with the real dumbbell interface samples shown in Table 1.

2.2 Experimental procedures

This experiment primarily discusses the emotional responses of the subjects to the interface software of the two groups of experimental equipment It integrates the feedback of the subjects to modify the questionnaire and the experiment content. The experimental process is as follows

(1) Testees: Three master's students from the engineering department and one interface expert were invited to perform experiments and give advice.
(2) Experimental devices 1: HTC Vive matched with a laptop was used to perform experiments with the self-developed VR motion software.
(3) Experimental devices 2: Real water dumbbell matching software was used for the experiment.
(4) After the experiment, the questionnaire was filled in and discussed with experts The questionnaire and the experiment content were then modified

2.3 Questionnaire framework

The questionnaire framework (divided into four parts) is shown in Table 2. The first part contains basic personal data The second part is the SAM scale for exploring emotional valence and arousal. The third part adopts QUIS to understand the interaction satisfaction of the interface. The fourth part presents the interview to understand the testees' intuitive feelings in using the process.

3 RESULTS AND DISCUSSION

3.1 Prototype of the development for a health incentive design

This project collected various types of data, learned and used Unity 3D to self-develop a health incentive software, and cooperated with HTC Vive headmounted display to conduct prototype simulation (Figure 1).

A detailed description of the prototype is presented as follows

1. Five game levels are offered for selection in the health incentive game.

Table 1. Sample description and characteristics: interface between VR motion software and real dumbbell.

Serial no.	1	2
Name	Active motivation	Entity detection
Picture		
Description	Perform personal challenges using VR applications	Use solid water dumbbells in personal challenges
Test site design		
Content	Software and intelligent virtual interactive equipment (virtual digital + VR device; active motivation model)	Simple physical equipment (physical equipment; passive incentive mode)
Provide tasks	1. Guidance and feedback on the completion of assigned tasks (dumbbell stretches) 2. Community communication design: Multi-person real-time activities or coaches to guide and cooperate with sports therapy 3. Design of incentive competition: Stretching completion competition, peer sports cumulative score competition	
Dimensions	Panoramic images	15 inch computer(s):
Degree of action	Low	High
Type	Digit	Entity
Guidance	Computer instruction guide	Real person coaching
Feedback	Sound effects and screen changes: Provides 3D stretching movement guidance, video and audio feedback, and 3D VR controller (balance simulated dumbbell) feedback	Coach interpretation: Provides human trainer stretching exercise guidance, oral feedback once exercise is achieved
Motivation types	Positive incentives / Negative incentives	Positive incentives / Negative incentives
Motivation content	DONE / Failed	DONE / Failed

2. Push the control button to pick up the dumbbell Select the menu. Follow the instructions for the point of contact, with both hands in the correct position to move to the next motion.

Table 2. Questionnaire content.

Questionnaire content	Number of questions	Question type	Purpose of questionnaire
Basic data	12 questions	Single choice	Understand personal background
SAM	2 questions	Likert scale	Emotional valence and arousal
QUIS	15 questions	Likert scale	Interface interaction
Advantages and disadvantages	3 questions	Questions of short description	Opinion

Figure 1. VR game screen design of health incentive software.

Table 3. Mean and standard deviation of emotional valence and arousal (the standard deviation is listed in the bracket, Likert scale: 1–5).

Group	VR application program	Real water dumbbell
Emotional valence	3.25 (0.96)	2.5 (1.00)
Emotional arousal	2.75 (1.26)	3.00 (0.82)

3. The same color dumbbell must touch the same color indicator.
4. In addition, positive excitation and negative excitation are designed. After the level is selected, a countdown is started. Once the task is completed within the assigned time, a completion pattern will appear. If the task is not completed in time, then a failure pattern will appear.

3.2 SAM scale

Comparing the two interfaces used by the testees, the statistical analysis results of the SAM scale are presented in Table 3 and Figure 2 for the differences between emotional valence and arousal.

The preceding results showed that the VR application programming interface is more pleasant The real water dumbbell interface is slightly stronger.

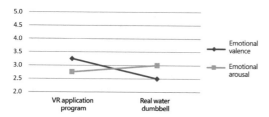

Figure 2. Comparison of emotional valence and arousal between the two interfaces.

Table 4. Mean and standard deviation of the QUIS scale (the standard deviation is listed in the bracket, Likert scale: 1–5).

Group	VR application program	Real water dumbbell
Overall response	3.38 (0.63)	3.06 (0.85)
Interface display	3.83 (0.52)	4.00 (0.50)
Interface information	2.92 (0.14)	3.08 (0.63)
Learnability	4.33 (0.52)	3.92 (0.76)

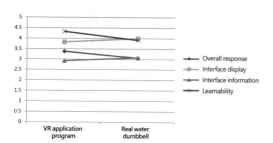

Figure 3. Comparison of the QUIS scale results of the two interfaces.

3.3 QUIS scale

In this study, the QUIS scale was used to compare the differences between the two interfaces in terms of overall user response, interface presentation, interface information, and interface learning. The statistical results are presented in Table 4 and Figure 3.

The preceding results showed that with regard to the overall response and learning of the interface, the VR applications are more positive than the real water dumbbells. However, with regard to interface presentation and interface information, the real water dumbbells are slightly more positive than the VR applications.

The overall statistics indicated that VR applications exhibited the highest learning behavior, but the real water dumbbells performed better in interface presentation. The interview revealed that VR applications provide simpler interface information that can be learned quickly. In addition, although the overall information of the interface is clear at a glance, the presented part gets a lower score due to style problem.

Table 5. Experimental and preliminary comparison results.

	VR fitness application program	Physical fitness equipment
Advantages	1. An instant response that can be designed (e.g., voice, image, animation) 2. Easy to understand image, and can also be required to provide feedback to the exact posture 3. Anchor points for changes are easier to design (design resources can be easily leveraged to expand) and can replace scenarios 4. Good positive excitation effect (sound and image animation)	1. With real weight 2. Adjustable weight 3. Easy to understand prompt screen 4. Free from hardware (starts immediately) 5. Easy to obtain props 6. High acceptance among the elderly
Disadvantages	1. It still requires testing for technology acceptance among the elderly. 2. It takes time to prepare (wearing and adjusting the equipment) 3. Starting requires additional calibrated height 4. Equipment is expensive and difficult to obtain 5. May cause dizziness due to personal adaptability (particularly for the elderly) 6. Too oppressive due to the close touching prompt 7. A drop in the weight sensitivity of the handle (dumbbell)	1. Manual guidance is still required 2. Additional mechanisms are necessary to ensure that the action is accurate. 3. No instant feedback 4. Less interesting 5. Less significance in terms of the incentive effect

3.4 Interview summary

The comprehensive interview results of this study are summarized in Table 5, and the details are described as follows.

The advantages of the VR application program include providing an instant response that can be designed (voice, image, or animation) and demonstrating good motivation effect. In the VR environment, the field of view is 110°, with good visual effect.

Meanwhile, one of the disadvantages of VR is that it still requires testing for technology acceptance among the elderly. VR takes time to prepare, and wearing and adjusting the equipment need the assistance of others. Therefore, it is insufficient for the real-time part.

Meanwhile, the advantages of the physical fitness equipment (i.e., water dumbbell) are its real weight real physical sense, and ability to achieve better exercise effect. Lastly, one of its convenient features is not being constrained by hard equipment.

Nevertheless, the disadvantages of the physical fitness equipment include the following: manual guidance is required and additional mechanisms are necessary to ensure that the action is accurate.

The aforementioned conclusions have been used as reference for the next experiment. In the future, VR interface modification will be performed using Unity software, and 30 testees will be recruited for the formal experiments. The results will then be statistically analyzed and presented in an accurate manner.

4 CONCLUSION

Compared with existing physical sports equipment, the advantages and disadvantages of the VR application developed in this study for the promotion of health among the elderly were identified. The preliminary conclusion of this study can be used as a reference for designers, and scholars are expected to conduct further explorations

In the future, VR interface modification is suggested to be performed using Unity software. More testees can also be recruited to conduct more accurate experiments.

REFERENCES

[1] Department of Economic and Social Affairs, United Nations Population Division. World Population Aging 1950–2050; Geneva: WHO, Department of Economic and Social Affairs Population Division (2002).

[2] Kobayashi, M.; Hiyama, A.; Miura, T.; Asakawa, C.; Hirose, M.; Ifukube, T. Elderly user evaluation of mobile touchscreen interactions. In Proceedings of the Human-Computer Interaction–INTERACT, Lisbon, Portugal, 5–9 September 2011; pp. 83–99.

[3] Bai, Y.W.; Chan, C.C.; Yu, C.H. Design and implementation of a simple user interface of a smartphone for the elderly. In Proceedings of the 2014 IEEE 3rd Global Conference on Consumer Electronics (GCCE), Tokyo, Japan, 7–10 October 2014; pp. 753–754.

[4] Marcus, A. Universal, Ubiquitous, User-Interface Design for the Disabled and Elderly. In HCI and User-Experience Design; Springer: London, UK, 2015, pp. 47–52.

[5] Fang Y-M, Lin C, Chu B-C. Older Adults' Usability and Emotional Reactions toward Text, Diagram, Image, and Animation Interfaces for Displaying Health Information. Applied Sciences. 2019; 9(6):1058.

[6] Chen, C. H., Kao, S. C., & Hu, M. C. (2018). Research on the Planning of Elderly Health Promotion Programs: Take the Learning Process of the Healthy Fitness Course as an Example. Journal of Adult and Lifelong Education, (30), 99–141.

[7] Fang, Y. M., Chang, C. C., (2016, Feb.) . Users' Psychological Perception and Perceived Readability of Wearable Devices for Elderly People. Behaviour & Information Technology, 35(3), 225–232.

[8] Tussyadiah, I.P.; Wang, D., Jung, T.H.; Dieck, M.C.T. Virtual reality, presence, and attitude change: Empirical evidence from tourism. *Tourism Management.* **2018,** 66, 140–154.

[9] Diemer, J.; Alpers, G.W.; Peperkorn, H.M.; Shiban, Y.; Mühlberger, A. The impact of perception and presence on emotional reactions: a review of research in virtual reality. *Frontiers in psychology.* **2015,** 6, 26.

[10] Schuemie, M. J.; Van Der Straaten, P.; Krijn, M.; Van Der Mast, C.A. Research on presence in virtual reality: A survey. *CyberPsychology & Behavior.* **2001,** 4(2), 183–201.

[11] Mehrabian, A., & Russell, J. A. (1974). An approach to environmental psychology. the MIT Press.

[12] Hodes, R. L.; Cook, E. W.; Lang, P. J. Individual differences in autonomic response: conditioned association or conditioned fear? Psychophysiology, 1985, 22(5): 545–560.

[13] Bradley, M. M.; Lang, P. J. Measuring emotion: the self-assessment manikin and the semantic differential. Journal of behavior therapy and experimental psychiatry, 1994, 25(1): 49–59.

[14] Chin, J. P.; Diehl, V. A.; Norman, K. L. Development of an instrument measuring user satisfaction of the human-computer interface. In: Proceedings of the SIGCHI conference on Human factors in computing systems. ACM, 1988. p. 213–218.

[15] Harper, B. D.; Norman, K. L. Improving user satisfaction: The questionnaire for user interaction satisfaction version 5.5. In: Proceedings of the 1st Annual Mid-Atlantic Human Factors Conference. 1993. p. 224–228.

[16] Tullis, T. S.; Stetson, J. N. A comparison of questionnaires for assessing website usability. In: *Usability professional association conference.* 2004. p. 1–12.

[17] ITRead01.com. The Exploratory Research of Unity and VR Industry – Applying Unity for VR. Source: https://www.itread01.com/content/1548778706.html (2015/05/15)

Smart Design, Science and Technology – Lam et al (eds)
© 2021 the Author(s), ISBN 978-1-032-01993-2

A new geared linkage mechanism with two length-variable links for path generation

Ren-Chung Soong*

Mechanical and Automation Engineering, Kao Yuan University, Kaohsiung, Taiwan

ABSTRACT: A new geared linkage mechanism with two length-variable links is proposed for planar path generation in this paper. This proposed mechanism consists of a five-bar linkage, a simple gear train, two elementary planetary gear trains and two slider-crank mechanisms. Two combinations, each one including an elementary planetary gear train, a slider-crank mechanism and a link adjacent to fixed link, serve as two length-variable driving links driven by a simple gear train. The proposed mechanism can be regarded as a geared five-bar linkage mechanism with two length-variable driving links. An optimal dimensional synthesis procedure is introduced to determine design variables for generating different type planar prescribed path trajectories. The formation, structure and kinematic analysis of the proposed mechanism are discussed in detail also. Two examples are provided to illustrate the proposed design method.

Keywords: Path generation, Geared linkage mechanism, Planetary gear train, Slider-crank mechanism, Kinematic analysis

1 INTRODUCTION

The control of a point in the plan to follow a pre-scribed path is called a planar path generation problem. In general, 1 DOF mechanisms, such as four-bar linkages, geared five-bar linkages, cam-linkages and cam-geared mechanisms are suggested to deal with this kind of path generation problems for considerations of simple structure and low cost.

Path generation problems of four-bar linkages have been studied deeply and extensively by different methods in past years. Zhou and Cheung (Zhou & Cheung 2001) proposed an optimal dimensional synthesis method based on the orientation structural error of the fixed link for path generation of crank-rocker linkages. Lin (Lin 2010) used an optimal dimensional synthesis method by applying a GA-DE hybrid evolutionary algorithm for path synthesis of four-bar linkages. Matekar and Gogate (Matekar & Gogate 2012) proposed an optimum synthesis method by using differential evolution and a modified error function for path generation of four-bar mechanisms. Peng and Sodhi (Peng & Sodhi 2010) presented an optimal synthesis method of approximate multi-phase path generation for adjustable mechanisms. Chanekar and Ghosal (Chanekar & Ghosal 2013) used an optimal synthesis method of adjustable planar four-bar crank-rocker mechanisms for approximate multi-path generation. Soong and Wu (Soong & Wu 2009) designed a 2-DOF four-bar linkage for generating multiple

types of coupler curves by controlling the angular displacement of the driving link and adjusting the link length of the fixed link. Zhou (Zhou 2009) synthesized an adjustable four-bar linkage for generating the precise continuous paths by adjusting the pivot location of the driven side link. Some important studies relate to path generations of geared five-bar linkage have been reported in past years. Buskiewicz (Buskiewicz 2010) proposed an optimal dimensional synthesis method based on shape invariants for path generations of geared five-bar linkage. Mundo et al. (Mundo et al. 2009) used a genetic algorithm method to synthesize optimal dimensions of five-bar linkage mechanisms driven by a pair of non-circular gears for generating exact prescribed paths. Lin (Lin 2015) presented an optimum synthesis method applied a combined discrete fourier descriptor for path generation of geared five-bar linkages. Some valuable research results with respect to path generation problems of the cam-linkage mechanisms were proposed by researchers in past decade. Ye and Smith (Ye & Smith 2005) presented an analytical method to trace a prescribed path for a combined cam-linkage mechanism with an oscillating roller follower. Mundo et al. (Mundo et al. 2006) proposed an optimal dimensional synthesis method of the cam-linkage mechanism with one or more disk cams for exact path generations. Gatti and Mundo (Gatti & Mundo 2007) proposed an optimal dimensional synthesis method of the cam-linkage mechanism with two disk cams for solving the problems of exact rigid-body guidance. Soong and Chang (Soong & Chang 2011) applied design concepts of inverse cam and

---

*Corresponding Author

DOI 10.1201/9781003188513-6

length-variable driving link to synthesize the optimal dimensions of cam-linkage mechanisms for exact function generations. Soong (Soong 2015, 2017) proposed a new simple design procedure for cam-geared mechanisms with the advantages of a simple and compact structure for the exact path generation and rigid body guidance problems. Gogate (Gogate 2016) proposed a design method of path generation based on inverse kinematic and dynamic analysis for planar adjustable mechanisms.

In this paper, a five-bar linkage, a simple gear train, two slider-crank mechanisms and two elementary planetary gear trains are combined to form a new 1 DOF geared linkage mechanism for solving problems of planar path generation. Formation, structure, kinematic analysis and dimensional synthesis procedure of the proposed mechanism will be discussed in following sections in detail.

2 FORMATION AND STRUCTURE

The new proposed planar mechanism is a 1-DOF geared linkage mechanism which is made up of a five-bar linkage, a simple gear train, two slider-crank mechanisms and two elementary planetary gear trains as shown in Figure 1. In this mechanism, Link 6 and Link 13 are fixed on Gear 2 and Gear 3 and rotate with them, respectively. Gear 1 and Gear 14 are fixed gears, therefore, the proposed mechanism degenerate as a twelve-bar geared linkage mechanism with twelve revolute joints, two sliding joints and four gear joints. The combination of Gear 1, Gear 4, Link 5, Link 6 and Slider 7 serves as one of length-variable driving links of the proposed mechanism. The combination of Gear 14, Gear 11, Link 13, Link 12 and the slider 10 serves the other one. In these two combinations, Gear 1 and Gear 4 have same pitch diameter, and Gear 11 and Gear 12 have same pitch diameter also. They are driven by a simple gear train which is made up of three same size gears, as Gear 2, Gear 3 and Gear 15 shown in Figure 1, respectively. Therefore, this new mechanism

can be regarded as a geared five-bar linkage with two length-variable driving links.

3 KINEMATIC ANALYSIS

The vector loop approach can be applied to conduct the kinematic analysis of the proposed mechanism in this section. The position vector loops and coordinate system of the proposed mechanism are shown in Figure 2.

According to the position vector loops shown in Figure 2, the vector-loop equation induced by $\vec{\rho}_6$, $\vec{\rho}_5$, $\vec{\rho}_4$ and $\vec{\rho}_2$ is

$$\vec{\rho}_6 - \vec{\rho}_5 - \vec{\rho}_4 - \vec{\rho}_2 = 0 \tag{1}$$

The scalar components of the expressions for Equation (1) on the X and Y axes are

$$\rho_6 \cos\theta_6 - \rho_5 \cos\theta_5 - \rho_4 \cos\theta_4 - \rho_2 \cos\theta_2 = 0 \tag{2a}$$
$$\rho_6 \sin\theta_6 - \rho_5 \sin\theta_5 - \rho_4 \sin\theta_4 - \rho_2 \sin\theta_2 = 0 \tag{2b}$$

Where ρ_i and θ_i are the magnitude and direction of vector $\vec{\rho}_i$ respectively, θ_6 is the angular position of driving Link 6, shown in Figure 1, and $\theta_2 = \theta_6 + 180°$, $\theta_4 = 2\theta_6 + \beta$, β is the initial direction of $\vec{\rho}_4$ corresponding to driving Link 6 in the horizontal position. Solving Equations (2) simultaneously, ρ_6 and θ_5, shown in Figure 2, can be obtained as follows:

$$\rho_6 = \frac{-b1 \pm \sqrt{b1^2 - 4c1}}{2} \tag{3}$$

$$\theta_5 = \tan^{-1}\left(\frac{\rho_6 \sin\theta_6 - \rho_4 \sin\theta_4 - \rho_2 \sin\theta_2}{\rho_6 \cos\theta_6 - \rho_4 \cos\theta_4 - \rho_2 \cos\theta_2}\right) \tag{4}$$

Where $b1 = 2\rho_2 - 2\rho_4 \cos\theta_6$, $c1 = \rho_4^2 - 2\rho_4\rho_2 \cos\theta_6 - \rho_5^2 + \rho_2^2$.

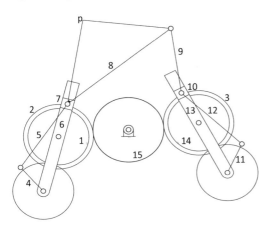

Figure 1. Formation and structure of new 1-DOF geared linkage mechanism.

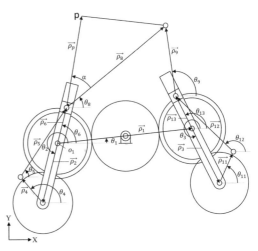

Figure 2. The position vector loops and coordinate system of the proposed mechanism.

According to the position vector loops as shown in Figure 2, the vector-loop equation obtained from $\vec{\rho}_{13}$, $\vec{\rho}_{12}, \vec{\rho}_{11}$ and $\vec{\rho}_3$ is

$$\vec{\rho}_{13} - \vec{\rho}_{12} - \vec{\rho}_{11} - \vec{\rho}_3 = 0 \tag{5}$$

The equations for the scalar components of Equation (5) on the X and Y axes are

$$\rho_{13}\cos\theta_{13} - \rho_{12}\cos\theta_{12} - \rho_{11}\cos\theta_{11} - \rho_3\cos\theta_3 = 0 \tag{6a}$$

$$\rho_{13}\sin\theta_{13} - \rho_{12}\sin\theta_{12} - \rho_{11}\sin\theta_{11} - \rho_3\sin\theta_3 = 0 \tag{6b}$$

Where θ_{13} is the angular position of Link 13 shown in Figure 1, $\theta_{13} = \theta_6 + \delta$ and δ is the initial direction of $\vec{\rho}_{13}$ corresponding to driving Link 6 in the horizontal position. $\theta_3 = \theta_{13} + 180°$, $\theta_{11} = 2\theta_6 + \gamma$, and γ is the initial direction of $\vec{\rho}_{11}$ corresponding to driving Link 6 in the horizontal position. Solving Equation (6) simultaneously, ρ_{13} and ρ_{12}, shown in Figure 2, can be obtained as follows:

$$\rho_{13} = \frac{-b2 \pm \sqrt{b2^2 - 4c2}}{2} \tag{7}$$

$$\theta_{12} = \tan^{-1}\left(\frac{\rho_{13}\sin\theta_{13} - \rho_{11}\sin\theta_{11} - \rho_3\sin\theta_3}{\rho_{13}\cos\theta_{13} - \rho_{11}\cos\theta_{11} - \rho_3\cos\theta_3} \right) \tag{8}$$

Where $b2 = 2\rho_2 - 2\rho_{11}\cos\theta_{13}$, $c2 = \rho_{11}^2 - 2\rho_{11}\rho_2\cos\theta_{13} - \rho_{12}^2 + \rho_2^2$.

According to the position vector loops shown in Figure 2, the vector-loop equation obtained from $\vec{\rho}_6$, $\vec{\rho}_8, \vec{\rho}_9, \vec{\rho}_{13}$ and $\vec{\rho}_1$ is

$$\vec{\rho}_6 + \vec{\rho}_8 - \vec{\rho}_9 - \vec{\rho}_{13} - \vec{\rho}_1 = 0 \tag{9}$$

The e equations for the scalar components of Equation (9) on the X and Y axes are

$$\rho_6\cos\theta_6 + \rho_8\cos\theta_8 - \rho_9\cos\theta_9 - \rho_{13}\cos\theta_{13} - \rho_1\cos\theta_1 = 0 \tag{10a}$$

$$\rho_6\sin\theta_6 + \rho_8\sin\theta_8 - \rho_9\sin\theta_9 - \rho_{13}\sin\theta_{13} - \rho_1\sin\theta_1 = 0 \tag{10b}$$

We can obtain the angular position of Link 9, (θ_9), and the instantaneous angular position of Link 8, (θ_8), by substituting ρ_6 and ρ_{13} into Equation (10) and solving the resulting two equations simultaneously:

$$\theta_9 = 2\tan^{-1}\left(\frac{-B \pm \sqrt{B^2 - C^2 + A^2}}{C - A} \right) \tag{11}$$

$$\theta_8 = \tan^{-1}\left(\frac{H}{G} \right) \tag{12}$$

Where

$A = 2\rho_1\rho_9\cos\theta_1 - 2\rho_6\rho_9\cos\theta_6 + 2\rho_{13}\rho_9\cos\theta_{13}$

$B = 2\rho_1\rho_9\sin\theta_1 - 2\rho_6\rho_9\sin\theta_6 + 2\rho_{13}\rho_9\sin\theta_{13}$

$C1 = 2\rho_1\rho_{13}(\cos\theta_1\cos\theta_{13} + \sin\theta_1\sin\theta_{13})$

$C2 = 2\rho_6\rho_{13}(\cos\theta_6\cos\theta_{13} + \sin\theta_6\sin\theta_{13})$

$C = \rho_1^2 + \rho_6^2 + \rho_9^2 + \rho_{13}^2 - \rho_8^2 - 2\rho_1\rho_6$
$\quad \times (\cos\theta_1\cos\theta_6 + \sin\theta_1\sin\theta_6) + C1 + C2$

$H = \rho_1\cos\theta_1 + \rho_9\cos\theta_9 + \rho_{13}\cos\theta_{13} - \rho_6\cos\theta_6$

$G = \rho_1\sin\theta_1 + \rho_9\sin\theta_9 + \rho_{13}\sin\theta_{13} - \rho_6\sin\theta_6$

Then, the position of the coupler point p on the X and Y axes can be obtained as follows:

$$p_x = o_{1x} + \rho_6\cos\theta_6 + \rho_p\cos(\theta_8 + \alpha) \tag{13}$$

$$p_y = o_{1y} + \rho_6\sin\theta_6 + \rho_p\sin(\theta_8 + \alpha) \tag{14}$$

Where o_{1x} and o_{1y} are the X and Y coordinates of fixed point o_1, respectively, and α represents the angle between the vectors $\vec{\rho}_p$ and $\vec{\rho}_8$, shown in Figure 2.

4 DIMENSIONAL SYNTHESIS

According to the formation, structure and results of our kinematic analysis of the proposed mechanism, synthesis problems with these dimensions require 18 design variables to be solved. In this section, an optimization procedure is applied to determine the design variables of the proposed mechanism for path generation problems. The general objective function for the optimization of dimensional synthesis can be defined as follows:

$$f(\rho_1, \rho_2, \rho_3, \rho_4, \rho_5, \rho_8, \rho_9, \rho_{11}, \rho_{12}, \rho_p, \\ \theta_1, \theta_{6ini}, \alpha_1, \beta, \gamma, \delta, o_{1x}, o_{1y}) = f_{obj} \tag{15}$$

The constrained equations of equality and inequality are also defined as

$$c_i(\rho_1, \rho_2, \rho_3, \rho_4, \rho_5, \rho_8, \rho_9, \rho_{11}, \rho_{12}, \rho_p, \\ \theta_1, \theta_{6ini}, \alpha_1, \beta, \gamma, \delta, o_{1x}, o_{1y}) = 0 \quad i = 1 \sim n_e \tag{16}$$

$$g_i(\rho_1, \rho_2, \rho_3, \rho_4, \rho_5, \rho_8, \rho_9, \rho_{11}, \rho_{12}, \rho_p, \\ \theta_1, \theta_{6ini}, \alpha_1, \beta, \gamma, \delta, o_{1x}, o_{1y}) < 0 \quad i = 1 \sim n_e \tag{17}$$

Where θ_{6ini} is the initial angular position of Link 6, which is shown in Figure 1, and f_{obj}, n_e and n_{ie} are the required objective function, the number of equality constrained equations and the number of inequality constrained equations, respectively.

5 DESIGN EXAMPLES

The following examples are provided to illustrate the design procedure. In general, the solution of a path generation problem is to design a mechanism for which the coupler point **p**, shown in Figure 2, minimizes the average structural errors of the generated path in comparison to the prescribed path. In this method, the average structural error is defined as:

$$error_{ave} = \sum_{i=1}^{n_p} \sqrt{(p_{xi} - p_{pxi})^2 + (p_{yi} - p_{pyi})^2}/n_p \tag{18}$$

Where p_{xi} and p_{yi} denote the X and Y coordinates of the ith position of coupler point p, respectively, p_{pxi} and p_{pyi} denote the X and Y coordinates of the position of the ith precision point in the prescribed path, and n_p is the number of precision points.

In terms of the X and Y coordinates of the precision points in the prescribed paths, the objective function and constrained equations are defined as follows:

Minimize

$$f(\rho_1, \rho_2, \rho_3, \rho_4, \rho_5, \rho_8, \rho_9, \rho_{11}, \rho_{12}, \rho_p,$$
$$\theta_1, \theta_{6ini}, \alpha_1, \beta, \gamma, \delta, o_{1x}, o_{1y}) = error_{ave} \quad (19)$$

Subject to the conditions for fully rotatable geared five-bar linkage that are reported in Ting (1994) and Li and Dai (2012).

The dimensions of ρ_1, ρ_6, ρ_8, ρ_9 and ρ_{12} shown in Figure 2, must satisfy the equality and inequality conditions imposed by the constraint equations for fully rotatable geared five-bar linkage defined in Ting (1994) and Li and Dai (2012). We determined the values of the design variables used in the following examples using the Fmincon function from the Matlab optimization toolbox.

5.1 EXAMPLE 1

In this example, the prescribed path is a closed path with a cusp and the 25 precision points with the coordinates listed in Table 1. The X and Y positions of all of the precision points along the prescribed and generated paths of the proposed mechanism are plotted in Figure 3. The values of the design variables are shown in Table 2. The link positions of the generated mechanism corresponding to the first precision point are shown in Figure 4.

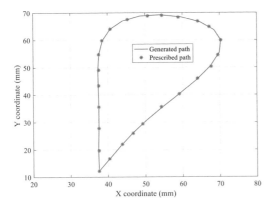

Figure 3. The generated and prescribed paths for Example 1.

5.2 EXAMPLE 2

An ellipse curve with 50 precision points was taken as the prescribed closed path in this example. The position of all precision points of the prescribed path and the generated curve of the proposed mechanism in terms of X and Y axes were shown in Figure 5. The valves of design variables were shown in Table 3. The link positions of the generated mechanism

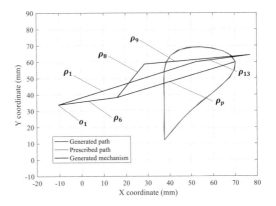

Figure 4. The link positions of the generated mechanism corresponding to the first precision point for Example 1.

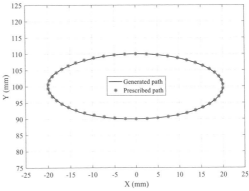

Figure 5. The generated path and prescribed path for Example 2.

corresponding to the first precision point were shown in Figure 6.

Since there were more precision points than design variables in two of our examples, the results show that the proposed mechanisms could generate approximate paths with minimal structural errors with respect to the prescribed paths, as expected. The dimensions of the mechanism generated in Examples 1 and 2 show that these two proposed mechanisms can be categorized as fully rotatable geared five-bar linkages of types B and A, respectively, as in Ting (1994) with two length-variable driving links. We assumed that all of the gear pairs in the proposed mechanism were of the same size, but plan to investigate path generation problems where the ratios between gear pairs vary in further studies. According to the discussion above, the proposed design approach could also be applied to solve problems of planar motion generation. Moreover, this method could be extended to the design of geared seven-bar linkages with translational or rotational output links for path, function and motion generation problems and industrial applications.

Table 1. Coordinates of the precision point for Example 1 (mm).

Point n	1	2	3	4	5	6	7	8	9	10	11	12	13
X-coordinate	70.3	67.2	64.1	58.9	54.4	50.7	45.3	40.7	38.5	37.6	37.6	37.6	37.6
Y-coordinate	59.9	64.8	66.8	68.3	69.0	68.8	67.5	64.0	59.8	54.7	49.1	43.4	35.6

Point n	14	15	16	17	18	19	20	21	22	23	24	25	26
X-coordinate	37.6	37.6	37.6	40.4	43.8	46.7	49.3	54.3	59.1	64.0	67.7	69.5	70.3
Y-coordinate	27.8	19.7	12.2	16.7	22.0	26.0	29.4	35.6	40.3	46.0	50.3	54.5	59.9

Table 2. Values of design variables for Example 1.

variables	ρ_1 (mm)	ρ_2 (mm)	ρ_3 (mm)	ρ_4 (mm)	ρ_5 (mm)	ρ_8 (mm)	ρ_9 (mm)	ρ_{11} (mm)	ρ_{12} (mm)
values	77.73	55.39	42.25	8.27	72.57	26.72	55.17	8.79	58.11

variables	ρ_p (mm)	θ_1 (°)	θ_{6ini} (°)	α (°)	β (°)	γ (°)	δ (°)	o_{1x} (mm)	o_{1y} (mm)
values	62.72	15.20	9.37	−28.63	3.19	−16.62	80.65	−14.81	34.06

Table 3. Values of design variables for Example 2.

variables	ρ_1 (mm)	ρ_2 (mm)	ρ_3 (mm)	ρ_4 (mm)	ρ_5 (mm)	ρ_8 (mm)	ρ_9 (mm)	ρ_{11} (mm)	ρ_{12} (mm)
values	18.74	169.01	117.39	18.28	23.45	170.03	156.37	54.72	155.50

variables	ρ_p (mm)	θ_1 (°)	θ_{6ini} (°)	α (°)	β (°)	γ (°)	δ (°)	o_{1x} (mm)	o_{1y} (mm)
values	142.51	36.35	−33.61	−45.96	65.21	35.06	−103.31	−8.90	115.51

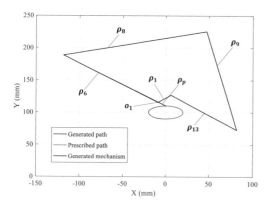

Figure 6. The link positions of generated mechanism corresponding to the first precision point for Example 2.

6 CONCLUSIONS

We have presented a new 1-DOF geared 12-bar linkage mechanism with two length-variable links for planar path generation. We composed the proposed mechanism using a five-bar linkage, a simple gear train, two elementary planetary gear trains and two slider-crank mechanisms. Two combinations, each including an elementary planetary gear train, a slider-crank mechanism and a link adjacent to the fixed link, serve as two length-variable driving links driven by a simple gear train. The proposed mechanism could be regarded as a geared five-bar linkage with two length-variable driving links. The fabrication, structure, coordinate system and kinematic analysis of the proposed mechanism were discussed in detail. We applied the Fmincon function from the Matlab software package to generate different types of prescribed planar paths. This optimal dimensional synthesis procedure was introduced to determine the values of the design variables. We illustrated the proposed design method by presenting two examples.

ACKNOWLEDGMENTS

The author is grateful to the Ministry of Science and Technology of the Republic of China (Taiwan, R.O.C.) for their support of this research under grant MOST 106-2221-E-224-004.

REFERENCES

Buskiewicz J., Use of shape invariants in optimal synthesis of geared five-bar linkage, Mechanism and Machine Theory, Vol. 45, No. 2 (2010), pp. 273–290.

Chanekar, P. V. and Ghosal, A., Optimal synthesis of adjustable planar four-bar crank-rocker mechanisms for approximate multi-path generation, Mechanism and Machine Theory, Vol. 69, (2013), pp. 263–277.

Gatti, G. and Mundo, D., Optimal synthesis of six-bar cammed-linkages for exact rigid-body guidance, Mechanism and Machine Theory, Vol. 42, No. 9 (2007), pp. 1069–1081.

Gogate, G. R., Inverse kinematic and dynamic analysis of planar path generating adjustable mechanism, Mechanism and Machine Theory, Vol. 102, No. 1 (2016), pp. 103–122.

Li, R. Q.; Dai, J. S., Workspace atlas and stroke analysis of seven-bar mechanisms with the translation-output, Mechanism and Machine Theory, Vol. 47, No. 1 (2012), pp. 117–134.

Lin, W. Y., A GA-DE hybrid evolutionary algorithm for path synthesis of four-bar linkage, Mechanism and Machine Theory, Vol. 45, No. 8 (2010), pp. 1096–1107.

Lin, W. Y., Optimum synthesis of planar mechanisms for path generation based on a combined discrete Fourier descriptor, ASME Transactions, Journal of Mechanism and Robotics, Vol. 7, No. 4 (2015), pp. 614–618.

Matekar, S. B. and Gogate, G. R., Optimum synthesis of path generating four-bar mechanisms using differential evolution and a modified error function, Mechanism and Machine Theory, Vol. 52 (2012), pp. 158–179.

Mundo, D., Lio, J. Y. and Yan, H. S., Optimal synthesis of cam-linkage mechanisms for precise path generation, ASME Journal of Mechanical Design, Vol. 128, No. 6 (2006), pp. 1253–1260.

Mundo, D., Gatti, G. and Dooner, D. B., Optimized five-bar linkages with non-circular gears for exact path generation, Mechanism and Machine Theory, Vol. 44, No. 4 (2009), pp. 751–760.

Peng C. and Sodhi, R. S., Optimal synthesis of adjustable mechanisms generating multi-phase approximate paths, Mechanism and Machine Theory, Vol. 45, No. 7 (2010), pp. 989–996.

Soong, R. C., A new cam-geared mechanism for exact path generation, Journal of Advanced Mechanical Design, Systems, and Manufacturing, Vol. 9 (2015), pp. 1–12.

Soong, R. C., A cam-geared mechanism for rigid body guidance, Transactions of the Canadian Society for Mechanical Engineering, Vol. 41, No. 1 (2017), pp. 143–157.

Soong, R. C. and Chang, S. B., Synthesis of function-generation mechanisms using variable length driving links, Mechanism and Machine Theory, Vol. 46, No. 11 (2011), pp. 1669–1706.

Soong, R. C. and Wu, S. L., Design of variable coupler curve four-bar mechanisms, Journal of the Chinese Society of Mechanical Engineers, Vol. 30, No. 3 (2009), pp. 249–257.

Ting, K. L., Mobility criteria of geared five-bar linkages, Mechanism and Machine Theory, Vol. 29, No. 2 (1994), pp. 251–264.

Ye, Z. and Smith, M. R., Design of a combined cam-linkage mechanism with an oscillating roller follower by an analytical method, Proc. IMechE Part C: Journal of Mechanical Engineering Science, Vol. 219, No. 4 (2005), pp. 419–427.

Zhou, H., Dimensional synthesis of adjustable path generation linkages using the optimal slider adjustment, Mechanism and Machine Theory, Vol. 44, No. 10 (2009), pp. 1866–1876.

Zhou, H. and Cheung, Edmund. H.M., Optimal synthesis of crank-rocker linkages for path generation using the orientation structural error of the fixed link, Mechanism and Machine Theory, Vol. 36, No. 8 (2001), pp. 973–982.

Smart Design, Science and Technology – Lam et al (eds)
© 2021 the Author(s), ISBN 978-1-032-01993-2

Research on the optimization of process parameters for MRR in PZT excited discharge channel compression Micro-EDM

LianMing Du* & XiangBo Ze
Department of Mechanical Engineering, Jinan University, Jinan, Shandong, China

QinHe Zhang
Department of Mechanical Engineering, Shandong University, Jinan, Shandong, China

ABSTRACT: Micro EDM is a kind of non-contact machining technology, which has the advantages of small macro cutting force and wide range of material application, but because of the precision of micro EDM, high discharge energy, small tool electrode size, easy to cause chip removal difficulties, resulting in unstable state of micro EDM, low machining efficiency and poor quality. In order to improve the environment and efficiency of micro EDM, this paper proposes the use of piezoelectricity based on micro EDM. A new EDM method of piezoelectric ceramic transducer (PZT) for synchronously excited compression discharge channel, which uses PZT when driving voltage of the actuator increases and compresses, the actuator will produce a small amount of elongation and retraction, which can effectively improve the state of the discharge gap, reduce the loss of the electrode, and increase the voltage of the actuator efficiency. In this paper, research on the optimization of process parameters for MRR in PZT Excited Discharge Channel Compression Micro-EDM is carried out by means of orthogonal test and signal-to-noise ratio analysis. The open circuit voltage, pulse width, pulse frequency and peak current are studied according to the large characteristic of SNR The SNR of material removal rate(MRR) is calculated, and the optimal parameter combination to achieve single machining target is obtained. The reliability of the optimization results is verified by experiments. The experiment shows that the open circuit voltage and pulse width are raised in the orthogonal experiment aiming at improving the material removal rate of workpiece. The material removal efficiency increases with the increase of open circuit voltage and pulse width, decreases first and then increases with the increase of pulse frequency, and increases with the peak value current increases first and then decreases.

1 INTRODUCTION

With the high-speed, miniaturization and precision of industrial products, micro EDM as a non-contact machining method is widely used in the machining of fine parts, molds and micro structures such as micro shafts and micro holes. Its processing range is widely used in aerospace, biotechnology, optical components, hydraulic and pneumatic products, fuel injection system and other fields (Bo et al. 2014, Li et al. 2019). Micro EDM is to use the spark discharge between electrode and workpiece to produce instantaneous high temperature to remove materials. The machining equipment is simple, the system is controllable, the macro cutting force is small, there is no contact stress between electrode and workpiece, and the range of Machinable materials is wide. It has unique advantages and broad application prospects in the field of micro machining technology, which can not be compared with other conventional machining technology.

The remarkable characteristics of micro EDM are small discharge gap, small discharge area, small single pulse discharge energy, small explosive spark, which is easy to cause chip difficult to machine, resulting in unstable discharge state, thus reducing machining efficiency and increasing electrode wear rate (Richard & Giandomenico 2018, Xiaohai et al. 2014, Zhang 2015, Zhu et al. 2010). In order to solve these problems, we use the converse piezoelectric effect of piezoelectric ceramic materials, and propose a new compression method of micro EDM based on the synchronous discharge channel excited by piezoelectric ceramic. By changing the discharge gap and compressing the discharge channel during the discharge process, we can promote the discharge of erosion products, so as to improve the steady-state discharge state of micro EDM.

2 EXPERIMENTAL

The experimental system of micro EDM with PZT excitation synchronous compression discharge channel is shown as Figure 1. PZT piezoelectric driver 2

---

*Corresponding Author

DOI 10.1201/9781003188513-7

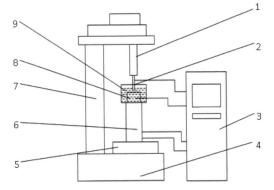

Figure 1. PZT excitation synchronous compression discharge channel micro EDM system diagram. 1 high-speed spindle 2 tool electrode 3. Sync pulse power supply 4. Granite base 5. X&Y table 6. PZT piezoelectric actuator 7. Z axis of the column 8 of the workpiece 9. Fluid

Table 1. Process parameters and level of orthogonal test.

Process parameters	Unit	Level		
		1	2	3
Open circuit voltage U_0	V	50	70	80
Pulse width t_{on}	μs	2	10	15
Pulse frequency f	kHz	6	8	9
Peak current I_m	mA	120	320	420

Table 2. The orthogonal test scheme with four factors and three levels.

Test No.	Open circuit voltage U_0 V	Pulse width t_{on} μs	Pulse frequency f kHz	Peak current I_m mA
1	50	2	6	120
2	50	10	8	320
3	50	15	9	420
4	70	2	8	420
5	70	10	9	120
6	70	15	6	320
7	80	2	9	320
8	80	10	6	420
9	80	15	8	120

is fixed on XY worktable, the maximum stroke in Z direction is 15 μm, and the resolution is 0.15 μm. The power supply 8 has two outputs: discharge circuit and piezoelectric actuator drive circuit, which drive pulse discharge spark and PZT piezoelectric ceramic feed at the same time. The two poles of the discharge circuit are respectively connected to the tool electrode 4 and the workpiece 7, and the Z axis drives the tool electrode to feed. When a certain gap is reached between the two poles, the working fluid is broken down. The driving circuit of piezoelectric driver is connected to the positive and negative ends of PZT driver to provide discharge pulse synchronous pulse signal and drive the driver to stretch. Open power pulse discharge voltage 0–120 v continuously adjustable pulse frequency 5–22 khz continuously adjustable, adjustable pulse width 1 μs to dozens of μs, peak current 50–500 mA adjustable.

In EDM, there are many factors that affect the test, such as open circuit voltage, discharge frequency, pulse width and peak current. These process parameters can be set at many levels, and they have certain rules in the single factor test, but in the actual test, the situation is more complex, and their mutual influence determines the machining of micro EDM Efficiency, electrode loss and surface quality.

As a new method of micro EDM, the orthogonal experiment design is used to study the processing technology of the synchronous compression discharge channel.

In the process of this test, the positive electrode is used for processing, the electrode is tungsten electrode, and the workpiece material is tool steel. See Table 1 for the processing parameters and levels selected in the orthogonal test.

3 RESULTS AND DISCUSSION

In addition to the influence of various electrical parameters, the machining effect of micro EDM technology is easily influenced by many other factors, such as the change of ambient temperature, the change of ambient humidity, the impurities in working fluid, the distribution of discharge points, etc., due to the small amount of micro EDM, the extremely fine tool electrodes and the small discharge gap It is easy to fluctuate. In this paper, the method of SNR analysis is used to analyze the results of orthogonal test.

The signal-to-noise ratio (s/N) analysis method can be used to analyze the influence of various test factors on the machining effect and consider the influence of noise signal. The signal-to-noise ratio is used as the measurement index of EDM effect in orthogonal test, which is convenient to find the combination of machining process parameters satisfying the ideal machining effect.

The characteristic indexes to measure the quality of micro EDM include: MRR, EWR and Ra. In an ideal state, the larger the MRR, the smaller the EWR, and the smaller the surface roughness Ra of the workpiece to be machined, it will be the better of the quality in EDM. There are two kinds of quality characteristic types in SNR analysis that correspond to our expected state: large characteristic SNR and small characteristic SNR. The calculation formulas are as follows:

$$S/N = -10 \lg \left(\frac{1}{n} \sum_{i=1}^{n} \frac{1}{y_i^2} \right) (db) \qquad (1)$$

$$S/N = -10 \lg \left(\frac{1}{n} \sum_{i=1}^{n} y_i^2 \right) (db) \qquad (2)$$

S/N — Signal to noise ratio of process target;
n — Number of test repetitions;
y_i — Process target value of the ith test.

Where (1) is the calculation formula of high the better (HB), and (2) is the calculation formula of low the better (LB). In the quality index of micro EDM, the material removal rate of workpiece has the expected characteristics. The larger the value is, the better the machining quality is. Therefore, formula (5-1) is used for calculation.

See Table 3 for orthogonal test results and signal-to-noise ratio analysis aiming at material removal rate of workpiece. See Table 4 for the average value of signal-to-noise ratio of workpiece material removal rate at all levels of processing parameters.

In Table 4, the range value r reflects the significance of the influence of different parameters on the material removal rate. The larger the range R of the mean value of the signal-to-noise ratio is, the more significant the influence of this parameter on the processing effect is, the greater the influence of the pulse width and open circuit voltage on the material removal rate is, while the influence of the peak current and pulse frequency on the material removal rate is smaller.

Figure 2 is the influence diagram of each factor on the mean value of signal-to-noise ratio of workpiece material removal rate.

Table 3. Orthogonal test results and signal-to-noise ratio analysis of workpiece material removal rate (MRR)

Test No.	MRR (10^{-6}mm^3/min)			S/N (db)
	Repeat test 1	Repeat test 2	Repeat test 3	
1	0.963470	0.909516	0.990293	0.421603
2	3.798384	4.054282	3.967955	11.900764
3	7.578270	8.558697	8.262719	18.170980
4	2.090345	2.703882	2.232167	7.240462
5	6.829076	4.048116	6.819826	14.602284
6	9.286310	9.631617	9.869016	19.633322
7	8.577196	8.638858	8.669689	18.718528
8	16.94166	12.10735	12.94595	22.656645
9	12.08885	12.01794	12.36941	21.695758

Table 4. Mean value of signal-to-noise ratio affected by each parameter on material removal rate.

Level	Open circuit voltage U_0	Pulse width t_{on}	Pulse frequency f	Peak current I_m
1	9.88	8.51	13.96	11.96
2	13.82	16.39	13.61	16.75
3	21.02	19.83	17.16	16.02
R	11.14	11.32	3.55	4.79

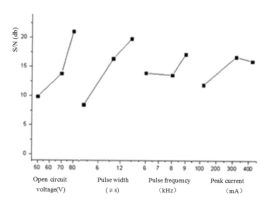

Figure 2. Influence diagram of each factor on the mean value of S/N ratio of workpiece MRR.

Table 5. Single objective optimization verification test results of workpiece MRR.

MRR (10^{-6}mm^3/min)			S/N (db)
Repeat test 1	Repeat test 2	Repeat test 3	
17.52436403	17.03723357	17.25921707	24.745931

It can be seen from Figure 2 that with the increase of open circuit voltage and pulse width, the signal-to-noise ratio of workpiece material removal rate increases gradually. With the increase of pulse frequency, the signal-to-noise ratio of workpiece material removal rate first decreases and then increases, while with the increase of peak current, the signal-to-noise ratio of workpiece material removal rate first increases and then decreases. When the signal-to-noise ratio of each factor is the maximum, the factor level combination can make the processing result of the optimized target reach the maximum signal-to-noise ratio, that is, the processing effect is the best, that is, when the open circuit voltage is taken as three levels 80 V, the pulse width is taken as three levels 15, the pulse frequency is taken as three levels 9 KHz, and the peak current is taken as two levels 320 mA, the maximum workpiece material removal rate can be obtained.

See Table 5 for the verification results of optimization test. Under the level combination of this factor, the signal-to-noise ratio of the material removal rate of the workpiece is 24.745931, which exceeds all the level combinations in Table 3 of the orthogonal test. The material removal rate of the workpiece is the largest, and the objective optimization result is reliable.

4 CONCLUSION

In this paper, based on the analysis of the influence of single factor on each processing index, the target optimization test of material removal rate index in processing quality index is carried out by using orthogonal test and signal-to-noise ratio analysis, and the conclusion is as follows:

Through the test and analysis of the influence of various processing factors on the material removal rate of workpiece, we can see that the open circuit voltage increases, the material removal rate of workpiece increases; the pulse width increases, the material removal rate of workpiece increases, the pulse frequency increases, the material removal rate of workpiece decreases first and then increases; the peak current increases, the material removal rate of workpiece increases first and then decreases.

REFERENCES

Zhu Bo, Zhong Xiaohong, Chen Jilun, Zhang Kun, et al., 2014, *Electromachining & Mould*, 3, 58–61.

Xiaopeng Li, Yuangang Wang, Yu Liu, et al. 2019, *Electromachining & Mould*, 19, 63–66.

Jacques Richard, Nicola Giandomenico, 2018, *Procedia CIRP* 68, 819–824.

Li Xiaohai, Liu Wuqi, Sun Zhaoning, Wang Xinrong, 2014, *Journal of Jiamusi University (Natural Science Edition)* 32, 899–900.

Zhang, Y., Xu, Z., Zhu, D., Xing, J., 2015. *Int. J. Mach. Tools Manuf.* 92, 10–18.

Zhu, D., Wang, W., Fang, X., Qu, N., Xu, Z., 2010. *CIRP Ann. – Manuf. Technol.* 59, 239–242.

Smart Design, Science and Technology – Lam et al (eds)
© 2021 the Author(s), ISBN 978-1-032-01993-2

Co-electrodeposition of manganese oxide/graphene electrodes for supercapacitors

Lung-Chuan Chen*, Guan-Ru Chen, Jean-Hong Chen & Yi-Chen Huang
Department of Materials Engineerring, Kun-Shan University, Tainan, Taiwan

ABSTRACT: A facial green co-electrodeposition of manganese oxide (MnOx)/reduced graphene oxide (RGO) supported on indium tin oxide (ITO) glass has been studied for supercapacitors. Graphene oxide (GO) is firstly synthesized from graphite by a modified Hummers method. Then the MnOx/RGO films were prepared by cyclic voltammetery (CV) from an aqueous mixture containing GO and manganese acetate. Characterization of the prepared MnOx/RGO electrodes were performed by XRD, FESEM, and FTIR. The capacitive behaviors of the MnOx/RGO electrodes were examined by CV, galvanostatic charge-discharge (GCD), and electrochemical impedance spectroscopy (EIS). The characterized analyses indicate that both manganese oxide and RGO can be successfully co-deposited on ITO glass. Manganese oxide is deposited in the morphology of nano particles and curved nano rods, while RGO is in flaky structure. Both MnO_2 and Mn_2O_3 without observable crystallinity are present in the deposited films. The Mn-O-C bonds are probable to exist in the samples. Appropriate modification of the manganese oxide films by co-electrodeposition of RGO can effectively improve their morphology, porosity, and electrical conductivity and reduce the electrolyte diffusion path, consequently enhancing the specific capacitance by about 100% at a scan rate of 100 mVs^{-1}.

1 INTRODUCTION

Supercapacitors (SCs) are important energy-storage devices as well as traditional capacitors and batteries. According to the charge-storage mechanism, SCs can be classified into two main categories, i.e., electrochemical double-layer capacitors (EDLCs) and pseudocapacitors (PCs). Owing to the property of excellent stability, low cost, long cycle time and strong corrosion resistance, EDLCs are the most widely used commercial supercapacitors fabricated with carbon materials, such as graphite, carbon nanotubes, and graphene as the electrode materials. In contrast, PCs are predominantly delivered by the electrodes constituted by transitional metal oxides, sulfides, and hydroxide, and conductive polymers. In general, EDLCs exhibits lower specific capacitance than PCs because of the absence of faradic redox reactions and morphology characteristics of carbonaceous materials, restricting charge accumulation in electrical double layer [1,2].

Manganese oxide has attracted much interest in acting as an active electrode material for supercapacitors because it is relatively cheap, non-toxic, and rich in resources, particularly, MnO_2 possesses ultra-high theoretical specific capacitance of ~1400 F/g [3, 4] due to the unique properties of multiple reversible variance states and broad potential window. However, the specific capacitance of the thick manganese oxide thin film electrode is quite low, and the actual specific capacitance is far from the theoretical value,

mainly due to the poor conductivity of manganese oxide ($10^{-5} \sim 10^{-6}$ Scm^{-1}), low specific surface area, low porosity and other factors [3–7]. Compositing MnOx with graphene has been reported to significantly increase the capacitive behavior because of the increase of specific surface areas and improvement of conductivity and morphology in addition to the introduction of non-faradic process. However, these procedures are usually complex and involve two separate steps or even more [8,9]. In this work, we employ a simple cyclic voltammetry technique to co-deposit reduced graphene oxide and manganese oxide on ITO glass. The prepared MnOx/RGO electrodes were characterized and their capacitance behaviors were examined by CV, GCD, and electrochemical impedance spectra (EIS), demonstrating that compositing graphene can significantly enhance the specific capacitance of MnOx.

2 EXPERIMENTAL

GO was prepared by a modified Hummers' method [10]. Typically, 1.5 g $NaNO_3$ was added into a solution containing 3.0 g graphite and 70 mL concentrated H_2SO_4. After which, the solution was cooled to 0°C in an ice bath, and then 9.0 g $KMnO_4$ was slowly added in portions with the temperature below 20°C under a vigorous stirring. Subsequently, the mixture was heated to 35~40°C and stirred for another 30 min. Once the solution turned into viscous gel type, 70 mL distillated water was added and stirred for 15 min at 98°C. The mixture was then cooled to room temperature and

---

*Corresponding Author

DOI 10.1201/9781003188513-8

was added by 20 mL of 30% H_2O_2 to make the mixture yellow. The resulting mixture was filtered, washed with dilute HCl solution and water. Finally, the obtained GO samples were dispersed into 500 mL ethanol with ultrasonication mixing.

The MnOx/RGOs were electrodeposited onto ITO glass by cyclic voltammetry. The ITO glass was cut into 10 mm × 30 mm and was ultrasonically rinsed in ethanol for 10 min. Before electrodeposition, the ITO glass was wiped with acetone and partly covered by polytetrafluorene ethylene films to make the exposed geometric area of 1 cm^2.

The electroplating solution consists of 2.319 g manganese acetate, 1.775 g sodium sulfate, and x mL of GO solution (x = 1,3,10) with water-ethanol (50:50 vol%) as the solvent. The samples with x of 1, 3, and 10 were denoted as GM1, GM3, and GMA, respectively. Electrodeposition of MnOx/RGO was performed in a three-electrode system with ITO, platinum plate, and Ag/AgCl as working, counter, and reference electrodes, respectively under a bias ranging from −1.4V to 1.4 V under a scan rate of 0.25 V/s. The electrodeposition lasted for 75 cycles (about 10 min), and then the prepared MnOx/RGO electrodes were thoroughly rinsed by distillated water, followed by drying in a 250°C oven for 2 h. For comparison, pure manganese oxide without RGO was also prepared and denoted as PM in this work.

The capacitive performance of the MnOx/RGO electrodes was evaluated by CV, GCD, and electrochemical impedance spectra (EIS) methods following a traditional three-electrode techniques using platinum and Ag/AgCl as counter and reference electrodes, respectively, on an AUTOLAB PGSTAT 320N Potentiostat (ECO Chemie BV).

Field emission scanning electron microscopy (FESEM) measurement was carried out on a JEOL JSM-7610F instrument. FTIR (Digilab) spectra were collected to reveal the bonding structures of the MnOx/RGO electrodes. The crystalline nature was examined by x-ray diffraction (XRD) with a PANalytical Empyrean diffractometer.

3   RESULTS AND DISCUSSION

The morphologies of the PM and GM3 samples are shown in Figure 1 with magnified times of 10k and 80k. The PM samples reveals homogeneous appearance at low magnification as shown in Figure 1a. When the magnification times increases to 80k, the surface of the PM sample is full of grains and curved nanorod shapes. The diameter and length of the nanorods range 15 to 35 nm and 15 to 100 nm, respectively. The RGO-modified GM3 sample displays uneven sheet and cracks under 10k magnification. Increasing the magnification time to 80k can clearly observe the existence of the nanorod structure as well as the sheet morphology. These sheet and nanorod appearances are ascribed to RGO and manganese oxides, respectively, indicating the co-electrodeposition of manganese oxides and RGO on ITO.

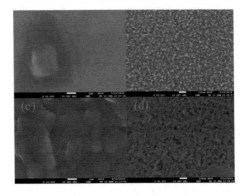

Figure 1.   FESEM images of the (a,b)PM and (c,d)GM3 electrodes.

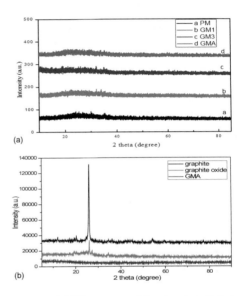

Figure 2.   (a) XRD patterns of the electrdeposited electrodes; (b) XRD patterns of the graphite, graphite oxide, and scratched GMA powder.

Figure 2a exhibits the XRD patterns of the prepared MnOx/RGO samples. The results imply no detectable MnOx crystalline, which may be arose from low crystallization, low content, and tiny diameter. The XRD of graphite and GO powders are also examined and show a predominant peak at 2θ of 26° and a tiny peak at 2θ of 20°. Moreover, the intensity of the diffractive peak of GO is significantly lower than that of graphite as shown in Figure 2b. This suggests that distance between layers of graphite is expanded during oxidation stage. Enlarging the XRD patterns of the GMA powder scratched from the electrode film exhibits some weak diffraction peaks at 2θ of 37.0°, 42.4°, and 56.1°, which are consistent with the (131), (300), and (160) crystalline plane of γ–MnO$_2$ (JCPDS 14-0644). Furthermore, a broad peak centered at 2θ around 7.0° can be obtained, which is probably resulted

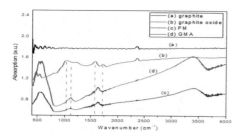

Figure 3. FTIR spectra of grphite, graphite oxide, and scratched PM and GMA powder.

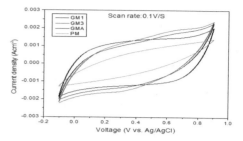

Figure 4. CV plots of the electrodeposited electrodes for capacitance measurement at 0.1 V/s.

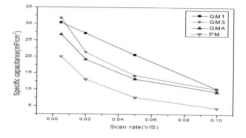

Figure 5. Specific capacitance with scan rate of CV tests.

from the RGO constitution. The shift of the diffraction angle from 20° to 7° also support the expansion of the layer distance, reflecting the reaction of GO to RGO during electrodeposition.

Figure 3 displays the FTIR spectra of the graphite, GO, PM, and GMA samples. Compared to graphite, the GO samples exhibit extra oxygen-containing groups of C-O ($1037\,cm^{-1}$), C-O-C ($1217\,cm^{-1}$), and C=O ($1722\,cm^{-1}$), supporting the oxidation of graphite. In addition, a band occurred at $1584\,cm^{-1}$ is considered to be caused by C=C bonds in the GO samples [11,12]. After compositing with MnOx, the band at $1722\,cm^{-1}$ in RGO samples disappears, which is likely due to the reduction reaction or modification of the carboxyl group of GO by MnOx during electrodeposition. In addition, the absorption band at $1584\,cm^{-1}$ becomes invisible in the GMA sample, which is suggested to be merged into the band at $1627\,cm^{-1}$ that caused by the adsorbed water molecules on the surface of MnOx/RGO samples. The broad absorption band distributed from 400 to $800\,cm^{-1}$ can be ascribed to the characteristic vibration of MnO_6^- octahedron in the PM and GMA samples, confirming the presence of MnOx [11,12]. In addition, the hydroxyl group contributes to the absorption band at $3400\,cm^{-1}$ for the PM and GMA samples. However, this band is very weak in the graphite and GO samples, indicating their poor hydrophilicity.

The capacitive behaviors of the prepared MnOx/RGO samples were assessed by CV and GCD techniques in a three-electrode configuration. Figure 4 depicts the CV results of the PM, GM1, GM3, and GMA samples in the potential range of $-0.1\,V$ to $0.9\,V$ (vs Ag/AgCl) at a scan rate of 0.1 V/s. The measurement with other scan rates of 0.005, 0.02, and 0.05 V/s were also examined (not shown here). The CV curves exhibit a nearly rectangular appearance without significant redox peaks for all the tested samples, indicating an ideal capacitive behavior. However, the CV curves seems to deviate from the rectangular shape and limit the potential window behind the tested range. The voltammetry current increases with the scan rate. A closer examination reveals that an insignificant redox pair occurs at 0.4 V and 0.6 V as the scan rate exceeds 0.05 V/s for GMA and GM3 samples, but this phenomenon is absent for the samples of PM and GM1. It is speculated that this redox reaction is due to the oxygen-containing group in RGO. The GM1 and PM

samples are lack of this redox peak because of the low content of RGO. The specific capacitance at various scan rates are summarized in Figure 4. At a low scan rate of 0.005 V/s, the specific capacitance decreases in order of GM3 > GM1 > GMA > PM. When the scan rates increase to 0.05 V/s or higher, the relation is GM1 > GM3 > GMA > PM. No matter the scan rate applied, the specific capacitance of the PM sample is the least, indicating incorporation of RGO can elevate the specific capacitance of MnOx. These results reflect that doping a proper amount of RGO in the manganese oxide can effectively improve the capacitive behavior of the MnOx/RGO composite. The reasons may be (1) the conductivity of RGO is higher than that of manganese oxide, therefore, doping RGO in manganese oxide can increase the electron transfer rate and effectively collect charges. (2) Doping a proper amount of RGO into manganese oxide can improve the pore structure of manganese oxide, facilitate the transportation of electrolytes in the pores, and reduce the diffusion resistance. (3) When the manganese oxide is dispersed as a nanostructure on the surface of the RGO sheet, the transmission distance of the electrolyte can be reduced, which is beneficial to the effective utilization of electrolytes.

Figure 6a exhibits the Nyquist plot of this system, and Figure 6b is the partial enlarged view of Figure 6a. As can be seen from Figure 6b, in a very high frequency region (100k–20k), a small arc impedance (R_1) is displayed at all four electrodes, which can be attributed to the effect of the electric double layer. Because the electric double-layer capacitor is mainly due to the double-layer arrangement of electrolyte ions at the interface between the electrode surface and the electrolyte, the electric double-layer process is not

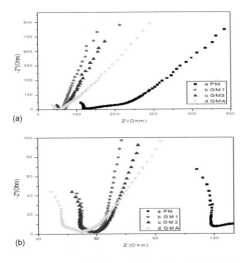

(a)

(b)

Figure 6. (a) Nyquist plots of the electrodeposited electrodes; (b) Partial enlargement of the Nyquist plots of the electrodeposited MnOx/RGO samples.

relevant to electrochemical reactions, therefore, the generation of R_1 is very fast and occurred in the very high frequency region [13,14]. Following the electric double-layer process is a shrinking semi-circular arc in the high-frequency region. The diameter of the semi-circular arc on the real axis represents the transfer resistance (R_2) of electrons between the electrode and the electrolyte, which is mainly related to the material's porosity and nanostructure. The left intersection of the semicircle and the real axis represents the equivalent series resistance or internal resistance (Rs) of the system, which is caused by the ionic conductivity of the electrolyte, the electronic conductivity of the electrode, the resistance of the carrier, and the contact resistance between the carrier and the active material [13–15]. The Rs of the prepared samples descends in order of PM > GM1 > GM3 > GMA. The PM sample contains no graphite oxide and therefore displays the lowest conductivity, so Rs is the largest among the tested samples. The GMA demonstrates the best conductivity because of the most doped graphite oxide, so that Rs is the smallest. On the other hand, the porosity of the GMA electrode may be the poorest among the three GO–containing samples, leading to the largest charge transfer resistance. The sudden increase of the impedance of the virtual axis in the intermediate frequency region indicates that the electrochemical system enters the capacitive behavior region, and the capacitive behavior becomes more significant as the frequency further decreases. The larger the slope of the curve, the closer to the ideal capacitive behavior. The impedance in the intermediate frequency region is predominantly due to the Warburg diffusion resistance, which is mainly related to the charge separation procedure within the micropores [15]. The impedance analyses are in good agreement with the specific capacitance phenomena.

4 CONCLUSION

The co-electrodeposition of MnOx and RGO has been successfully fabricated through a CV method. The MnOx were electrodeposited on ITO as nano particles or curved nano rods, while RGO were mainly electrodeposited as sheet appearance. Appropriate doping RGO is beneficial to improve the pore structure and morphology of the MnOx/RGO films, which is in favor of electrolyte's diffusion. The electrodeposited films show an amorphous structure, while the powder scraped off from the film has weak crystallinity. Doping a suitable amount of RGO in MnOx can improve the conductivity, pore structure, and reduce the transported distance of electrolytes, causing the enhancement of the specific capacitance of the MnOx/RGO composite electrode.

ACKNOWLEDGMENT

We acknowledge the financial support provided by the Southern Taiwan Science Park Bureau, Ministry of science and Technology, Taiwan, R.O.C. under grant 108CE02.

REFERENCES

[1] G. Wang, L. Zhang, J. Zhang, Chem. Soc. Rev., 2012, 41, 797–828.
[2] M. Serrapede, A. Rafique, M. Fontana, A. Zine, P. Rivolo, S. Bianco, L. Chetibi, E. Tresso, A. Lamberti, Carbon, 2019, 144, 91–100.
[3] H. Zhou, X. Zou, Y. Zhang, Electrochimica Acta, 2016, 192, 259–267.
[4] P. R. Jadhav, M.P. Suryawanshi, D.S. Dalavi, D.S. Patil…, Electrochimica Acta, 2015, 176, 523–532.
[5] Q. Chen, Y. Meng. C. Hu, Y. Zhao, H. Shao, N. Chen. L. Qu. J. Power Sources. 2014, 247, 32–39.
[6] Z. Hu, X. Xiao, C. Chen, T. Li, L. Huang, C. Zhang, J. Su, L. Miao, J. Jiang, Nano Energy, 2015, 11, 226–234.
[7] H.-M. Lee, S.-W.Cho, C.J. Song, H. J. Kang, B.-J. Kwon, C.-K. Kim, Electrochimica Acta, 2015, 160, 50–56.
[8] Q. Chen, Y. Meng. C. Hu, Y. Zhao, H. Shao, N. Chen. L. Qu. J. Power Sources. 2014, 247, 32–39.
[9] W. K. Chee, H. N. Lim. Z. Zainal, N. M. Huang, I. Harrison, Y. Andou, J. Phys. Chem. C, 2016, 120, 4153–4172.
[10] D.C. Marcano, D.V. Kosynkin, J.M. Berlin, A. Sinitskii, Z. Sun, A. Slesarev, L.B. Alemany, W. Lu, J.M. Tour, ACS Nano 2010, 4, 4806–4814.
[11] Z. Li, Y. Mi, X. Liu, S. Liu, S. Yang, J. Wang, J. Mater. Chem. 2011, 21, 14706–14711.
[12] G.S. Gund, D.P. Dubal, B.H. Patil, S.S. Shinde, C.D. Lokhande, Electrochimica Acta 2013, 92, 205–215.
[13] C.-C. Hu, C.-C. Wang, Journal of Electrochemical Soc. 2003, 150, A1079–A1084.
[14] W. Wei, X. Cui, W. Chen, D.G. Ivey, Electrochimica Acta 2009, 54, 2271–2275.
[15] A. Di Fabio, A. Giorgi, M. Mastragostino, F. Soavi., Journal of the Electrochemical Society 2001, 148, A845–A850.

Smart Design, Science and Technology – Lam et al (eds)
© 2021 the Author(s), ISBN 978-1-032-01993-2

The influence of the international workshop teaching mode on students' learning effectiveness and interaction

Syuan-Lan Shih*, Chun-Kuan Wu, Chung-Shun Feng, Tsu-Wu Hu, Sheng-Jung Ou &
Ming-Yu Hsiao
Department of Industrial Design, Chaoyang University of Technology, Wufeng, Taichung, Taiwan

ABSTRACT: According to previous research, using workshop teaching methods in the design field can generate different benefits, such as enhancing creativity, knowledge, social interaction, and overall design performance. This research is based on the theme of "eco-smart life design," established a week-long international seminar, and finally tried to produce product or service design. In other words, the purpose of this study is to explore the effects of learning and interaction between students of different cultural backgrounds by using design studios as teaching models. According to the literature review, learning effects include objective performance evaluation and subjective learning achievements. In order to achieve the research purpose, the research used quantitative and qualitative methods. First, this study uses questionnaire surveys and expert discussions to assess students' learning outcomes. Second, the study used qualitative interviews to collect data about student interaction status. After the investigation, the research results were analyzed using descriptive statistics, difference tests and content analysis. The results show that students generally take a positive attitude towards international workshops as a good teaching model and are willing to participate in similar activities again. Regarding student interactions, I have received much interesting feedback about lifestyle habits, work attitudes and seminar content. Finally, the study suggests that follow-up studies can be conducted in different countries. By using the same teaching structure for more research, replacing local teaching resources, and comparing the learning and interactive effects of students from different national backgrounds and local resources, it is expected that the benefits and operational details of the design seminar can be further confirmed.

Keywords: International workshop; Learning outcomes; Industrial design; Cross-cultural

1   INTRODUCTION

The workshop is a kind of method which can create more learning effect than that of the general teaching method in the study. The design workshop was introduced which focused on the advantages of green production and marketing, intelligent digital technology, cultural innovation, and industrial marketing. In the process, students should follow the concept of the striving for the savings in resource use and the reduction of pollution. Students were suggested to use the intelligent technology and integrated data to enhance the effectiveness and efficiency of innovation. Students can also conduct surveys and analyses from a variety of issues and propose design alternatives from various perspectives. This design pattern can also be called "issue design" which is to present a problem in the world today. Through the guidance of the workshop teaching model, students can discuss their solutions, enhance their thinking ability through training, and raising their attention to social issues.

2   LITERATURE REVIEW

2.1   *International design workshop*

Whether in academia or industry, the working mode of the design workshop has been implemented in the field of industrial design for a long time. It was initiated by an American company named IDEO about 20 years ago, and it is still in wide use. The actual execution process of the workshop does not follow a standard guideline. Through cooperation and competition, industry peers such as Acer, Phillips, and IDEO, advocators of the application of scenario, have gained some of their creativity and inspiration from utilizing this workshop (Huang 2006). [1] Taiwan has been implementing this approach for several years. For example, Chinese Industrial Designers Association has jointly organized the ADA International Student Design Workshop with its Japanese and South Korean counterparts 17 times since 2002. Due to the intensity, timeliness and flexibility of this kind of action pattern in the workshop, different arrangements can be made according to various conditions to achieve

---

*Corresponding Author

                                                    DOI 10.1201/9781003188513-9

distinct desired results–a fact that suits the "integrated" model (Yao 2002). [2] The purpose of the workshop is to gather staff and students from different backgrounds to implement various design methods, which may involve the commonly used brainstorming, KJ method, conceptual sketch, design situation, etc. The subsequent work mode would hopefully involve mutual stimulation and mutual assistance. Particularly, implementing these methods have the most obvious effects on international workshop. Due to the different cultural backgrounds of members, the experience and opinions on a single topic also vary a great deal. Ultimately, this diversity has resulted in the most effective use of brainstorming.

We must realize that it is difficult for University programmers to keep up with the pace at which technical knowledge is advancing. Therefore, we need a different kind of teaching, more flexible and adaptable (Ramón Rubio García 2014). [3] Therefore, to break through the constraints of the current university curriculum, one must be willing to try a new teaching model.

2.2 Learning outcomes

Generally, the definition of learning outcomes refers to the improvement of knowledge, skills, and attitudes after students received the education. It is an indicator for measuring learning outcomes and one of the most important categories in teaching quality evaluation (Hsu 2010). [4] It's seen as a basis for learners and educators to improve and adjust in the future. In addition, Hong (1999) considered that the specific learning effects should be divided into (1). Objective learning outcomes, including test scores, completion schedule time and semester grades; (2). Subjective learning gains, including learning satisfaction, achievement, and preferences, etc. and consider the objective assessment and the feelings of the learners themselves. [5] Objective learning outcomes generally refer to actual achievements, and subjective learning outcomes refer to psychological accomplishment and satisfaction.

Regarding the objective learning effect, Lin (2012) also mentioned five indicators of relevant learning outcomes, including: (1) Temporality – after a period of study or training; (2) Acquired – through acquired learning; (3) Measurability – to measure the extent to which learning outcomes are achieved according to prior goals; (4) Content – focusing on skills or knowledge in a particular field; and (5) Diversity – the expression of learning outcomes are all different in different subjects. [6] Chen (2003) believes that during the learning process, learning satisfaction refers to the extent to which learners achieve their own expectations and needs in the process of learning. [7]

In brief, we can see that learning outcomes refer to the degree to which learners gain satisfaction on the objective results of the grades as well as the subjective expectations of their skills or knowledge enhancement according to relevant reviews or tests after the learners have accepted the education for a period of time.

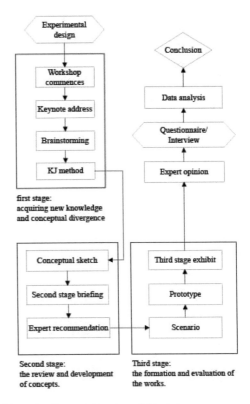

first stage:
acquiring new knowledge and conceptual divergence

Second stage:
the review and development of concepts.

Third stage:
the formation and evaluation of the works.

Figure 1.   Research flowchart of this study.

3   RESEARCH METHOD

This study attempts to explore the effects on learning produced by the interaction between the teaching method of the International Design Workshop and the students. There are three main groups of students: students from Philippines' Mapua University, students from Japan's Shibaura Institute of Technology and students from Taiwan's Chaoyang University of Technology. These three types of students all major in industrial design, are close in age, and have the same level of expertise. A total of 40 students, mixed with different nationalities, are divided into 6 groups that are guided by six specialized teachers. The contents of the workshop are divided into three stages: student learning and the dissemination of ideas, the review and development of concepts and the formation and review of proposal. After the workshop has ended, the use of questionnaire surveys and personal interviews would enable the students to conduct self-assessments and voice their feelings. The design of the questionnaire is divided into three parts: communication, work, and "inner feelings." The main thing to be evaluated is "the subjective learning effects," As for the objective learning effects, six experts with relevant experience and background, would evaluate the final works. The obtained scores are ranked as "objective learning results." The research architecture flowchart is shown in Figure 1.

Figure 2. Students organized generated ideas via KJ methods.

Figure 3. Students sketched their ideas for six experts to evaluate.

Figure 4. Final presentation with prototype and scenario.

4 EXPERIMENT OPERATION

4.1 The content of workshop

The theme of the workshop is "ECO-Smart Living Design." It is divided into three stages as mentioned above. The first stage involves acquiring new knowledge and conceptual divergence. Two experts are invited to give two keynote speeches, which is followed by brainstorming, KJ methods, and so on to organize generated ideas as shown in Figure 2. The second stage represents the review and development of concepts. Students sketch their ideas for six experts to evaluate as illustrated in Figure 3. In the third stage, the formation and evaluation of the works were implemented. The establishment of the prototype and its usage scenario would enable students to discuss and make changes as needed. The final study was the final presentation. It would allow six experts to critique and discuss with the students, as shown in Figure 4.

4.2 Questionnaire and interview

The content of the questionnaire is divided into three parts. The first part mainly deals with communication. The questions include: 1. I think there is no difficulty in communicating with people from different countries; 2. I think the problems with communication stem

Table 1. Score of each question in this study.

Questionnaires	Mean	Standard deviation
I think there is no difficulty in communicating with people from different countries.	1.90	0.966
I think the problems with communication stem from language barriers.	3.35	1.099
I think there are cultural barriers to effective communication.	2.88	1.159
I think I have a good working relationship with people from different cultural backgrounds	3.25	1.256
I think people from different countries may have distinct operating mode.	3.8	0.966
I think people from different countries may have distinct work pace.	3.88	0.966
working with people from different cultural backgrounds makes me work harder than before	3.58	1.196
I like this working mode.	4.18	0.813

from language barriers; 3. I think there are cultural barriers to effective communication. The second part discusses work. The questions involve: 1. I think I have a good working relationship with people from different cultural backgrounds; 2. I think people from different countries may have distinct operating modes; 3. I think people from different countries may have distinct work paces. The third part concerns people's inner feelings. The questions are: 1. working with people from different cultural backgrounds makes me work harder than before; and 2. I like this working mode. The questionnaire employs a Likert scale of 1 to 5. Additionally, the interview focuses on the entire workshop experience and how it feels to work with group members from various cultural backgrounds.

In total, 40 valid questionnaires were obtained. Using an independent sample t-test, one notices that students from the Philippines, Japan and Taiwan have no significant differences in their answers to these three major parts. As for the overall averages, in a communication-related question, "I think there is no difficulty in communicating with people from different countries," the average number is 1.9 on the Likert scale. Such a low value likely demonstrates that students generally feel that communicating effectively is difficult, but still think highly of this type of operating mode as shown in a high average score of 4.18. Table 1 illustrates the average score of each question. Furthermore, one of the questions "if there is a similar workshop in the future, I would like to participate," only one student answers negatively, while the remaining 39 answer in the affirmative.

In the interview, most of the students have mentioned cultural impacts on their lives. For example,

6 Taiwanese students have indicated that Japanese students seem more austere, while Filipino students were more outgoing, and this attribute is reflected in their work. Hence, Filipino students are better at divergent thinking, while Japanese students have a superior ability in materializing their ideas. Concerning communication skills, 11 students from all 3 countries have mentioned that it is their first time working with members from different cultural backgrounds. Although facing an initial difficult adjustment period, they are able to produce great ideas and results soon afterwards. Since the venue is in Taiwan, Filipino and Japanese students have shared some observations about living in Taiwan for the past week. For example, Filipino students have been surprised by the orderly sorting of household waste in Taiwan, while Japanese students have been impressed by the diversity of night markets and streets.

5   CONCLUSION

The experimental results show that the participants, regardless of nationality, generally believe that although a considerable degree of difficulty in communication exits, most of them say they like the teaching model of the international workshop, but are mindful of the fact that the cultural impact is so strong to the point of affecting their relationship with colleagues. On the whole, the results of the workshop are very good. In less than a week, the subjective learning experience, objective learning evaluation, and the special cultural learning experience brought to students is even more valuable.

Future research directions will be doing similar studies in different countries. One would employ the same teaching structure but replace the local teaching resources. In this way, one could compare the effects on student learning and interaction with respect to different national backgrounds and resources.

REFERENCES

[1] Huang, C. F., A Study of Experiential Scenario Design Workshop on the Industrial Academia Cooperation -the case study on the Cheerful, National Taipei University of Technology, 2006.
[2] Yao, C. W., Experiential Scenario Design – A Study of the Design Relay Workshop of AURORA and SENAO Co-Project, National Taipei University of Technology, 2002.
[3] Ramón Rubio García, Eco-design workshop: Delivering new concepts for future designs, Gijon Polytechnic School of Engineering University of Oviedo, 2014.
[4] Hsu. L. C., A Study of Skill Competitor' Involvement, Learning Motivation and Learning Effectiveness— A Case of Textual Skill Processing for High School, Chaoyang University of Technology, 2010.
[5] Hong, M. Z., *Teaching on Internet*. Publisher: Soft china, 1999.
[6] Lin, P. W., A Study of the Relationships among Skill Learning Styles, Self-regulated Learning Strategies and Learning Effectiveness of Students – A Case of Vocational High School Food and Beverage Management Department in Kaohsiung. National Kaohsiung University of Applied Sciences, 2012.
[7] Chen, Y. M., the Effects of Learning Style and Learning Mode on the Effectiveness of e-Learning in Junior High Schools National Chung Cheng University, 2003.

Smart Design, Science and Technology – Lam et al (eds)
© 2021 the Author(s), ISBN 978-1-032-01993-2

The impact of hardware buffer size settings on digital audio production: The model example of the Avid pro tools digital audio workstation

Ching-Chien Liang & Chao-Chih Huang*
Department of Popular Music Industry, Southern Taiwan University of Science and Technology, Yungkang, Tainan, Taiwan

Chian-Fan Liou
Department of Visual Communication Design, Southern Taiwan University of Science and Technology, Yungkang, Tainan, Taiwan

ABSTRACT: With the developments of audio workstation, the functions of software are becoming more powerful for engineers. In terms of audio processing, the related engineering, as well as combination with video productions and the parameter settings, the hardware buffer size should be modified in order to keep the production flows smooth in a timely manner. Nowadays, most audio engineering has long relied on software to perform nonlinear processing. In the technical specifications of audio workstation, there are two roles relied for computing: the Central Processing Unit (CPU) and Digital Signal Processing (DSP) cards. In the past, due to the stronger computing power of the DSP card, it was relied for the main computing, while CPU was used for the auxiliary computing work. Recently, due to the processing power of the CPU becoming more powerful, budgeted studios may then choose not to purchase expensive DSP cards, under the considerations of the cost. In audio production, the status of monitoring latency is the most affected issue by the hardware buffer size during recording. After recording, the issue affected by hardware buffer size is the processing power of the CPU left for plug-ins processing during mixing. This paper takes the model example of the world-famous Avid Pro Tools audio workstation to explore the impact of hardware buffer size settings, and performs in-depth interviews, descriptive statistics and linear statistical analysis to investigate the psychological differences toward engineers. The findings of this research are expected to suggest better hardware buffer size settings for audio production.

1 INTRODUCTION

With the rapid evolution of digital audio hardware and software, the functions of audio recording, editing and post-production software and hardware are becoming more powerful for both consumers and professional engineers. In terms of digital audio processing techniques, the related audio recording, editing, mixing and mastering engineering, as well as the subsequent combination output with audio and video productions which may involve digital video data, the session parameter settings in digital audio software, the hardware buffer size settings, must be properly modified and adjusted, to keep the production tasks smooth.

In the technical specifications of workstation for digital audio production, there are two roles relied for computing: the Central Processing Unit (CPU) and the Digital Signal Processing (DSP) card. Nowadays, under the considerations of the cost for purchasing audio workstation, due to the processing power of the host computer, and due to the CPUs becoming more and more powerful, small and medium-sized budget recording studio owners may choose not to purchase expensive DSP cards in the early start-up of their business. This is a new trend in that CPU can be used for replacing the DSP card to do the main audio computing tasks.

In the process of audio production, the status of recording monitoring latency is the most direct affected issue by the hardware buffer size settings during actual analog audio recording. After the audio recording, the issue affected by hardware buffer size settings is the processing power of the host computer CPU left for processing real-time and non-real-time audio plug-ins during editing and mixing work.

This research takes the model example of the world-famous Avid Pro Tools Digital Audio Workstation system to explore the impact of hardware buffer size settings. The findings of this research are expected to suggest better hardware buffer size parameter settings for audio recording, editing, mixing and mastering processes.

2 EXPERIMENTAL

In recent academic studies, scholars have had proposed relevant research on digital audio workstation, such as composing (Marrington 2017) and music production

---

*Corresponding Author

DOI 10.1201/9781003188513-10

Table 1. Different hardware buffer sizes used for surveys.

No.	Hardware Buffer Size
1	256 Samples
2	512 Samples
3	1024 Samples

(Stables et al. 2016). However, there are not many studies focused on the impacts caused by the settings of the hardware buffer sizes within digital audio workstations. If appropriate research can be developed with the modifications of the hardware buffer sizes, the research results will help professional and consumer audio engineers with the responsive logic for the preparation and operation work.

This study is based on descriptive statistics (Libman 2010; Marshall & Jonker 2010) and linear statistical analysis (Li et al. 2016) and explores the psychological differences toward engineers and three sets of parameter settings of different hardware buffer sises. The data shown in Table 1 was applied in the Avid Pro Tools digital audio workstation to check for the psychological differences toward different respondents.

In this research, 45 university students, who had participated in music production tasks for more than 5 projects in recording studios, with ages between 18 to 21 years old, were invited as the respondents. A four point Likert Scale questionnaire, as shown in Table 2, was designed for each hardware buffer size parameter setting to ask the respondents to reply with their opinions about the impressions of the audio playback latency level, under different hardware buffer size parameter settings. The audio equipment settings and the interview steps were as follows:

(1) The control variables were all under the same conditions: all the respondents took the interviews in the same recording studio, using the same microphone (RØDE, NTR), same microphone preamplifier (Midas, XL48), same audio-to-digital digital-to-audio converter interface (Avid, HD I/O), same computer (Mac Pro, 2.7 GHz 12-Core Intel Xeon E5 processor, 64 GB memory), same headphones (AKG, K240) and the same monitor speakers (Digidesign, RM1, PMC designed speakers).

(2) Avid Pro Tools digital audio workstation was used with the interview procedures (version: Pro Tools | Ultimate 2020.3).

(3) The audio recording session parameter settings were set to: (a) bit depth: 16-bit, (b) sample rate: 44.1 kHz, (c) record audio file type: Broadcast Wave Format (BWF), (d) record track type: one single mono audio track.

(4) The respondents were asked to sing or do oral speaking into the microphone and properly wear the headphone to listen to the audio playback signal relayed by the microphone to and from the studio control room and booth.

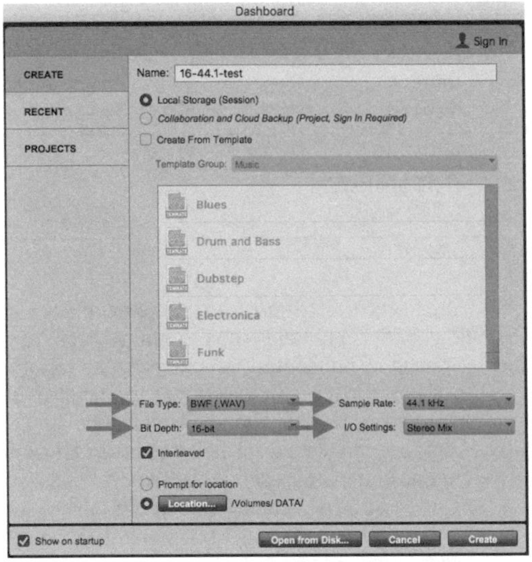

Figure 1. Pro Tools session parameter settings screen. (Courtesy of Avid Technology, Inc., www.avid.com.).

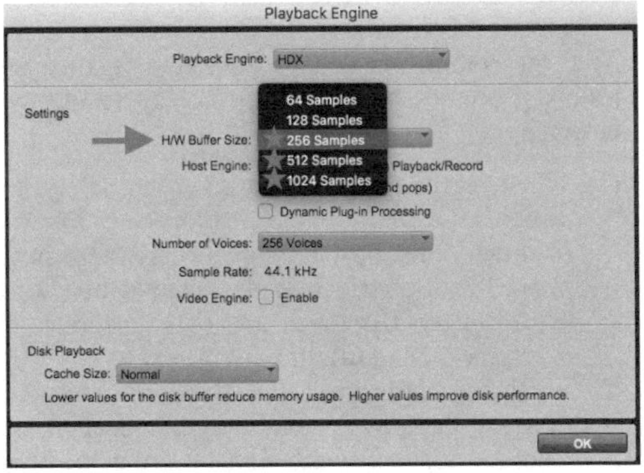

Figure 2. Pro Tools hardware buffer size settings screen. (Courtesy of Avid Technology, Inc., www.avid.com.).

Figure 3. Pro Tools session under recording screen. (Courtesy of Avid Technology, Inc., www.avid.com.).

41

Table 2. Four point Likert Scale questionnaire.

No.	Hardware Buffer Size	Strongly Disagree	Disagree	Agree	Strongly Agree
1	256 Samples				
2	512 Samples				
3	1024 Samples				

Table 3. Different hardware buffer sizes' lagged time in theory.

No.	Hardware Buffer Size	Lagged Time
1	256 Samples	0.0058 second
2	512 Samples	0.0116 second
3	1024 Samples	0.0232 second

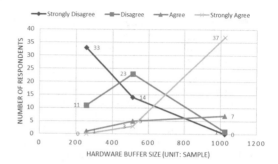

Figure 4. Responses to different hardware buffer sizes.

(5) Three sets of hardware buffer size parameter settings, 256, 512 and 1024 samples, were used for each respondent as the independent variables to check for the opinions about the level of the impressions to the audio playback latency.

(6) All 45 respondents were asked to finish checking the questionnaire as shown in Table 2. The statement question presented in the questionnaire was: "the recorded audio playback signal heard in the headphones *is* lagged."

3 RESULTS AND DISCUSSION

After the completion of the above investigation and testing processes, the following survey results was obtained:

(1) When the hardware buffer size was set to 256 samples, the answer counts to the question "the recorded audio playback signal heard in the headphones *is* lagged," from 45 respondents, were: (a) Strongly Disagree: 33, (b) Disagree: 11, (c) Agree: 1, and (d) Strongly Agree: 0.

(2) When the hardware buffer size was set to 512 samples, the answer counts, to the question "the recorded audio playback signal heard in the headphones *is* lagged," from 45 respondents, were: (a) Strongly Disagree: 14, (b) Disagree: 23, (c) Agree: 5, and (d) Strongly Agree: 3.

(3) When the hardware buffer size was set to 1024 samples, the answer counts, to the question "the recorded audio playback signal heard in the headphones *is* lagged," from 45 respondents, were: (a) Strongly Disagree: 0, (b) Disagree: 1, (c) Agree: 7, and (d) Strongly Agree: 37.

This research was based on the settings of the audio recording session parameters set to bit depth at 16-bit and sample rate at 44.1 kHz, thus in theory, when hardware buffer size was set to 256 samples, the recorded audio playback signal heard in headphone would be lagged for 0.0058 second; when hardware buffer size was set to 512 samples, the audio playback signal heard in headphone would be lagged for 0.0116 seconds, and inferentially, when hardware buffer size was set to 1024 samples, the recorded audio playback signal heard in headphone would be lagged for 0.0232 second, which reached the maximum audio monitoring latency under the three sets of hardware buffer parameter settings in this research.

Based on the three sets of survey data results listed above, this research sorted out the experimental presentation results in Figure 1 that 37 out of 45 respondents replied when hardware buffer size was set to 1024 samples, the recorded audio playback signal heard in headphone was lagged seriously; whilst 33 out of 45 respondents replied when hardware buffer size was set to 256 samples, they strongly disagree that the recorded audio playback signal heard in headphone was lagged. As for when hardware buffer size was set to 512 samples, the answers from the respondents were not extremely as tended to "strongly disagree" or "strongly agree" as the two sets of answers with 256 or 1024 samples.

4 CONCLUSION

In this paper, three different hardware buffer size parameter settings were used to check respondents' feedback opinions to the level of audio playback latency. As for the results of the questionnaire surveys in this research, and the calculations of three different audio playback latency time values based on the premise of audio sample rate set to 44,100 samples per second, it was confirmed that the higher the hardware buffer size parameter settings, the greater the level of audio playback monitoring latency will occur during live audio recording.

Regarding the study of audio playback situations, experiments and investigations of live analog audio recording for musicians or singers in recording studio were implemented in this research. As for the practical work in recording studio, one common audio playback task is, audio recording engineers simply playback the existing sound data that has been recorded or stored in digital audio workstation; in other words, simply to perform the "play" function.

Each manufacturer's digital audio workstation has its own ways to activate the "play" function. For Avid Pro Tools digital audio workstation, one of the several ways to simply activate the "play" function is just to press the spacebar of the computer keyboard directly. After pressing the spacebar of the computer keyboard, the computer system will start to read the existing digital audio data in the digital audio workstation, then convert the digital audio data to analog audio signal, and finally the audio recording engineers, musicians or singers can hear the sound with audio monitors or headphones.

Between the moments from pressing the spacebar of the computer keyboard until audio recording engineers are capable to hear the sound with audio monitors or headphones, under different hardware buffer size parameter settings theoretically, the latency times will be different. This theoretical logic is similar to the situations of live analog audio recording for musicians or singers in recording studio: the higher the hardware buffer size parameter settings are, the greater the latency time differences will be; the lower the hardware buffer size parameter settings are, the less the latency time differences will be. For the psychological expectations of the following two tasks, live analog audio recording for musicians or singers in recording studio, and audio recording engineers simply playback the existing sound data that has been recorded or stored in digital audio workstation, the latency time difference in the former task, when the audio playback is heard, should be in strict conditions to be expected as small as possible by listeners; whilst the latency time difference when the audio playback is heard, may or may not be strictly high demanded by listeners in the latter task

The reasonable inference from this is that the position of the playback start point at the digital audio workstation computer screen does not necessarily have to be precisely located within any audio waveform amplitude portion in every audio playback action, thus the playback start area might include silence, space or nearly silenced unimportant noise. Therefore, in terms of psychological expectations of listeners, for the task of audio recording engineers simply playing back the existing sound data that has been recorded or stored in digital audio workstation, the hardware buffer size parameter settings will not cause the listeners to experience obvious audio playback latency time differences as the task of live analog audio recording for musicians or singers in the recording studio.

This research checked three different levels of audio playback monitoring latency under three different hardware buffer size parameter settings in the process of analog audio live recording, suggesting that in order to reduce the recording monitoring latency to the minimum, when musicians or singers are doing live analog audio recording, the hardware buffer size parameter settings should be set as low as possible; and during the subsequent process of audio editing and mixing, this research suggests that the hardware buffer size parameter settings can be set to higher the better without making listeners in studios experience serious audio playback latency.

One of the common recording studio tasks, Musical Instrument Digital Interface (MIDI) recording, was not experimented and discussed in this research to explore the impact of hardware buffer size parameter settings to MIDI recording capabilities. For future areas of research suggestion, the existing discourses in MIDI recording topic, it was mentioned in the Avid Pro Tools reference guide document that lower hardware buffer size parameter settings are useful for improving latency issues in certain recording situations or system performance problems: On Pro Tools systems, lower settings reduce MIDI-to-audio latency (when playing a virtual instrument live and monitoring the instrument)(Avid Technology, Inc. [Avid] 2020). It is suggested that more research on how and what the levels of different hardware buffer size parameter settings will cause and affect digital audio workstation systems' MIDI performances in various kinds of certain recording situations.

REFERENCES

Avid Technology, Inc. (2020). *Pro Tools® Reference Guide Version 2020.9* (p. 94). Burlington, MA: Avid Technology, Inc.

Li, L., Du, L., Zhang, W., He, H., & Wang, P. (2016). Enhancing information discriminant analysis: Feature extraction with linear statistical model and information-theoretic criteria. *Pattern Recognition*, 60, 554–570.

Libman, Z. (2010). Alternative assessment in higher education: An experience in descriptive statistics. *Studies in Educational Evaluation*, 36(1–2), 62–68.

Marrington, M. (2017). Composing with the Digital Audio Workstation. In J. Williams & K. Williams (Eds.), *The Singer-Songwriter Handbook* (pp. 77–89). New York, NY: Bloomsbury Academic.

Marshall, G., & Jonker, L. (2010). An introduction to descriptive statistics: A review and practical guide. *Radiography*, 16(4), e1–e7.

Stables, R., Man, B. D., Enderby, S., Reiss, J. D., Fazekas, G., & Wilmering, T. (2016). Semantic Description of Timbral Transformations in Music Production. *MM '16: Proceedings of the 24th ACM international conference on Multimedia*, pp. 337–341.

Smart Design, Science and Technology – Lam et al (eds)
© 2021 the Author(s), ISBN 978-1-032-01993-2

Optimization design and verification of milling insert geometry for chamfering process

Devendra Reddy Bandi* & Shinn-Liang Chang
Department of Power Mechanical Engineering, National Formosa University, Hu-Wei, Yunlin, Taiwan

Cheng-Hsiung Chen
General Manager, Chain Headway Machine Tools Co., Ltd., Taiwan

ABSTRACT: Precision tool manufacturing companies always assume to have their products with high precision to compete in the market. This requires constant improvements in the tool design. The aim of this study is to optimize the tool design of the milling insert for the chamfering process. Chamfering is a basic process applied in any machining industry for the deburring or removal of sharp edges of the work material. The main aspect of a chamfered surface is to have a good surface finish. This study investigates the effect of insert geometry on the chamfered surface finish by the use of Taguchi L9 orthogonal array design. Control factors considered in this study are the rake angle, the land width, and the honing radius. All the experiments were conducted with two sets of process parameters as per Taguchi design. The process parameters used are 6000 rpm spindle speed, 400 mm/min feed rate for the chamfer size of C2, and 200 mm/min for the chamfer size of C3. The results were found to be that insert geometry with the rake of 18°, land width of 0.2 mm, and honing radius of 20 μm produces a chamfered surface with a good surface finish.

1   INTRODUCTION

Chamfering is a machining operation in which sharp edges of the component are eliminated by a slope cut at any right-angled edge, which greatly eliminates the chance of cuts and injury to individuals engaged with the processing of metal pieces. Chamfering also makes the assembly smooth and easy [1]. The essential aspect of a chamfered surface is good surface finish. Surface finish quality plays a major role in today's manufacturing market. From the customer's viewpoint, quality is very important because the extent of quality determines the degree of satisfaction of the customers [2].

Wang et al. [3] studied the effect of cutting-edge preparation techniques on cutting edge strength & properties. The results showed that drag finishing has good performance than other techniques in terms of tool life, cutting force generation, lower flank wear, and better surface roughness on the finished product. Nareen et al. [4] studied the effect of process parameters on surface generation for aluminum alloy. The results in his study showed that the surface roughness generated on the machined part would be lower at

higher spindle speeds and lower feed rates. Although the insert geometry, process parameters, and cutting-edge preparation techniques will have an impact on the surface generated, due to the available resources, this study did not examine all of these effects. This study is limited to investigate the effect of insert geometry parameters such as rake angle, land width, and honing on surface roughness in the chamfering of S50C steel with TiAlN, PVD-coated, TCEX13T304E solid carbide insert.

This study aims to find the best insert geometry parameter combination which will produce a chamfered surface with good surface finish, to replace the existing insert.

2   METHODOLOGY

Experiments are designed with the help of Taguchi L9 orthogonal array (Figure 1). The software used for DOE (Design of Experiments) is Minitab 19. This study is divided into four main phases, those are planning phase, test insert manufacturing phase, experimentation phase or conduction phase, and analyzing phase.

---

*Corresponding Author

DOI 10.1201/9781003188513-11

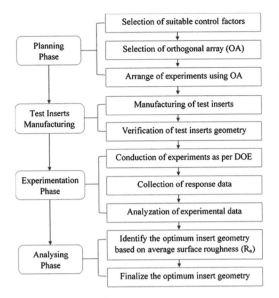

Figure 1. Research methodology

3 PLANNING PHASE

3.1 Selection of input parameters

Table 1. Selection of parameters.

Control factor	Symbol	Units
Rake Angle	Factor-A	(°)
Land Width	Factor-B	mm
Honing Radius	Factor-C	μm

Table 2. Existing insert geometry.

Control factors		Units
Rake Angle (A)	15	(°)
Land Width (B)	0.24	mm
Honing Radius (C)	34	μm

Table 3. Selection of parameter level.

	Factor Levels		
Control Factors	L1	L2	L3
Rake Angle (A)	12	15	18
Land Width (B)	0.15	02	025
Honing Radius (C)	20	30	40

3.2 Design of experiments

For the selected input parameters, Taguchi L_9 orthogonal array design was used to arrange the inserts that were manufactured. Taguchi orthogonal array is a fraction design that aims to provide optimum results at the best possible experimental test cost by eliminating the excess sample test that need to be performed in the event of full factorial design [5]. For this purpose, software Minitab 19 was used. Table 4 shows the inserts to designed and manufactured.

Table 4. Design of experiments.

Test Insert	Rake Angle °	Land Width (mm)	Honing Radius (μm)
1	12	0.15	20
2	12	0.2	30
3	12	0.25	40
4	15	0.15	30
5	15	0.2	40
6	15	0.25	20
7	18	0.15	40
8	18	0.2	20
9	18	0.25	30

3.3 Workpiece material

The material used for the experiment is JIS S50C, whose dimensions are $130 \times 100 \times 44$ mm ($l \times b \times h$) respectively. Workpiece composition is shown in table 5 [6].

Table 5. Workpiece chemical and mechanical properties.

Element	Composition
C	0.47–0.53 %
Mn	0.60–0.90 %
P	0.030 %
S	0.035 %
Si	0.15–0.35 %
Tensile strength	630 MPa
Yield strength	375 MPa

4 TEST INSERT MANUFACTURING PHASE

Any production process begins with the design and the idea of how to produce the designed model into the finished product. The test inserts to be manufactured were designed using solidworks, figure 2 shows the 3D model of existing insert. Similarly, all test inserts were designed by adjusting rake angle, land width, and honing radius as per Table 4. In order to reduce the production cost only one cutting edge is manufactured.

The production process involves the planning of the necessary manufacturing processes, tools, fixtures, etc. There may be a number of manufacturing processes for a single feature, but in a cost-effective way,

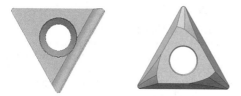

Figure 2. Existing insert 3D model.

Figure 3. Illustration of insert manufacturing process.

it is important to choose the best possible process with less time and cost. According to the available resources, the following manufacturing processes are selected for manufacturing of TCEX13T304E:

1. Raw material preparation
2. Peripheral grinding process
3. Wire EDM (Electrical Discharge Machining) process
4. Cutting edge preparation process
5. Insert naming process
6. Coating process

Figure 3 depicts the manufacturing processes involved in the test inserts production.

5 EXPERIMENTATION AND DATA COLLECTION

5.1 Conduction phase

After manufacturing of test inserts, all the documented experiments were conducted in two phases. The machining center used during this study is DAHLIH BA-1020 milling machine. Process parameters used in this study are shown in table 6. After every experiment the chamfered surface was tested for its surface roughness using Mahr surface roughness measuring device and ARCS yuantai 2D surface imaging machine.

Table 6. Process parameters.

| Parameters | Units | Value | |
		Set 1	Set 2
Spindle Speed	r.p.m	6000	6000
Feed Rate	mm/min	400	200
Chamfer Size	mm	C2	C3

In this study, 10 inserts (9 test inserts and one existing insert) were tested for two sets of process parameters. Initially, all the inserts were tested for first set of process parameters and response data were collected. Based on the response data only few inserts were selected for second phase testing.

5.2 Equipments used in this study

1. Chamfering Cutter

The chamfering cutter and insert used in this study are SSP-1616-110L (SSP 45° Spot Chamfering Cutter) and TCEX13T304E-AR. These are the products of the Chain Headway Machine Tools Co., Ltd. Tool holder and collet used for holding the chamfering cutter are BT50-PNER32, PNER32-16 [7]. Tool holder and collet are the products of SYIC company. Figure 4 shows the tool combination of the chamfering cutter.

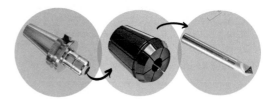

Figure 4. Tool combination - chamfering cutter

2. Surface Roughness Measuring Instrument

The device used to measure surface roughness in this study is Perthometer M1, a product of Mahr. A pair of magnetic V-blocks were used to hold the workpiece at 45° to measure the surface roughness of the chamfered surface. Figure 5 shows the Mahr's surface roughness measuring device.

Figure 5. Perthometer-M1.

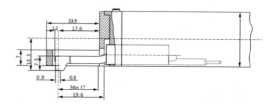

Figure 6. Perthometer-M1 stylus.

46

Table 7. Perthometer-M1 specifications.

Measuring principle	stylus method
Traversing speed	0.5 mm/s
Measuring ranges	100 μm
Profile resolution	12 nm
Filter	Gaussian
Cut-offs	0.25/0.8/2.5 mm
Traversing lengths as per DIN/ISO	1.75/5.6/17.5* mm
No. of sampling lengths	1.25/4/12.5* mm
Standards	DIN/ISO/JIS/ASME
Parameters	Ra, Rz, Rmax, RPc JIS: Ra, Rz
Dim. (L × W × H)	190 mm × 170 mm × 75 mm

Table 8. Response data for first set of process parameters.

Test no	Rake angle	Land width	Honing radius	Spindle Speed	Feed Rate	Depth of cut	Surface Roughness			Raw Edge quality
							R_{a1}	R_{a2}	Ra avg	
Unit	(°)	mm	mm	rpm	mm/min	mm	μm	μm	μm	quality
1	12	0.15	20				0.877	0.712	0.7945	2
2	12	0.2	30				0.626	0.604	0.615	2
3	12	0.25	40				0.253	0.225	0.239	2
4	15	0.15	30				0.292	0.264	0.278	3
5	15	0.2	40	6000	400	C2	0.218	0.207	0.2125	1
6	15	0.25	20				0.871	0.571	0.721	3
7	18	0.15	40				0.509	0.490	0.4995	2
8	18	0.2	20				0.185	0.163	0.174	2
9	18	0.25	30				0.317	0.357	0.337	2
10	Existing Insert						0.332	0.511	0.4215	2

Figure 6 illustrates the detailed view of the stylus, and Table 7 shows the specifications of the Perthometer-M1 [8].

3. Yuantai ARCS 2D Imaging Machine

The Yuantai 2D imaging machine is used to check the 2-dimensional geometry of any object within minutes. In this study, it is used to inspect the chamfered surface generated by all the inserts. Figure 7 shows the 2D imaging process of insert surface [9].

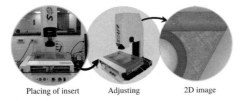

Placing of insert	Adjusting	2D image

Figure 7. ARCS 2D imaging process.

Table 9. Response data for second set of process parameters.

Insert No.	Rake Angle	Land Width	Honing Radius	Spindle Speed	Feed Rate	Depth of Cut	Surface Roughness			R. E
							Ra1	Ra2	Ravg	
Unit	°	mm	mm	rpm	mm/min	mm	μm	μm	μm	quality
3	12	0.25	40				0.313	0.574	0.4435	2.5
4	15	0.15	30				0.353	0.262	0.3075	2.5
5	15	0.2	40	6000	200	C3	0.647	0.624	0.6355	2
8	18	0.2	20				0.248	0.257	0.2525	2
10	Existing Insert						0.914	0.702	0.808	2

5.3 Data collection

After all the experiments, the response data were collected and tabulated in two phases. The data collected during phase one and phase two were listed in Table 8 and Table 9 respectively. Figure 8 shows the process of experiment conduction and data collection.

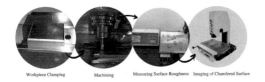

Workpiece Clamping	Machining	Measuring Surface Roughness	Imaging of Chamfered Surface

Figure 8. Experimentation and data collection.

After the first phase experimentation, From the Table 8, it can be observed that surface roughness of insert three, insert four, insert five, insert eight, and insert nine are less than the average surface roughness of existing insert.

Figure 9. (a) Chamfered surface generated by existing insert (b) 2D image of chamfered surface at focal length 1.0 (c), (d) 2D images of the chamfered surface at focal length 2.0.

Based on the response data inserts 3, 4, 5, 8 were selected for second phase experimentation. The second phase experimentation were conducted in order to cross verify the best available insert geometry which produces good chamfer surface.

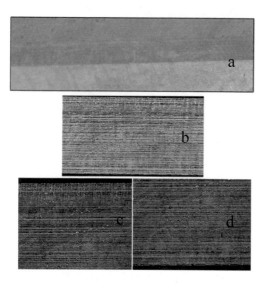

Figure 10. (a) Chamfered surface generated by insert 3 (b) 2D image of chamfered surface at focal length 1.0 (c), (d) 2D images of the chamfered surface at focal length 2.0.

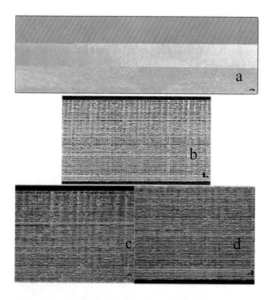

Figure 11. (a) Chamfered surface generated by insert 4 (b) 2D image of chamfered surface at focal length 1.0 (c), (d) 2D images of the chamfered surface at focal length 2.0.

6 ANALYSIS OF EXPERIMENTAL DATA

6.1 Conduction phase

Figure 9 to figure 13 reveals that the chamfered surface generated by the existing insert has a lot of cutting marks in the feed direction. It also can be witnessed from Table 8 that the surface roughness of the chamfered surface generated by existing insert is 0.4215 μm, which is higher than the surface roughness generated by the inserts three, four, five, eight, and nine.

From figure 13 and Table 8 it can also be observed that the chamfered surface produced by insert 8 is smoother, with almost no feed marks and the average surface roughness generated by insert eight is 0.174 μm, which is the least average surface roughness compared to surface roughness generated by other inserts. By comparing surface roughness data from Table 8 and surface produced from figures 9 $\sim$ 13, it can be clearly understood that insert eight is the best alternative for existing the insert.

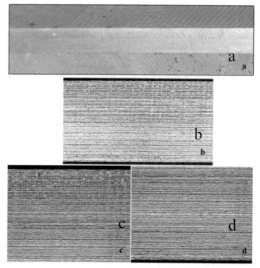

Figure 12. (a) Chamfered surface generated by insert 5 (b) 2D image of chamfered surface at focal length 1.0 (c), (d) 2D images of the chamfered surface at focal length 2.0.

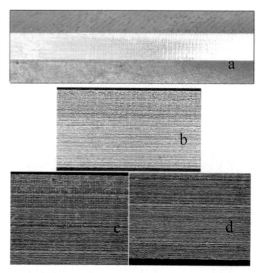

Figure 13. (a) Chamfered surface generated by insert 8 (b) 2D image of chamfered surface at focal length 1.0 (c), (d) 2D images of the chamfered surface at focal length 2.0.

7 CONCLUSION

In this study, nine test inserts with different geometry parameters were designed, manufactured and tested with the existing insert for the chamfering process of S50C to select the best insert that produce a better chamfered surface.

By observing the experimental results, the following conclusions were made:

1. Among nine test inserts manufactured, the insert number eight with a rake angle of 18°, land width of 0.2 mm, and honing radius of 20 μm produced the chamfer surface with a least average surface roughness Ra = 0.174 μm.
2. From these results, it can be considered that insert eight can be used as the best alternative for existing insert.

ACKNOWLEDGEMENTS

The work outlined in this paper was financially supported by MOST109-2221-E-150-011 project.

REFERENCES

"Wikipedia," [Online]. Available: https://en.wikipedia.org/wiki/Chamfer. [Accessed 2020].

J. Pratyusha, U. Ashok kumar, P. Laxminarayana, "Optimization of process parameters for Milling using Taguchi Methods," International Journal of advanced Trends in computer science and Engineering, vol. 2, no. 6, pp. 129–135, 2013.

Wanting Wang, M.D.Khalid Saifullah, Robert Abmuth, Dirk Biermann, A.F.M Arif "Effect of edge preparation technologies on cutting edge properties and tool performance," The International Journal of Advanced Manufacturing Technology, pp. 160; 1823–1838, 2020.

Nareen Hafidh Obaeed, Mostafa Adel Abdullah, Momena Muath, Maryam Adnam, Hind Amir "Study The Effect of Process Parameters of CNC Milling Surface Generation Using Al-alloy 7024," Diyala Journal of Engineering Sciences, vol. 12, no. 03, pp. 103–112, 2019.

J. M. Cimbala, "Taguchi Orthogonal Arrays," Peen State University, 2014.

"ASTM Steel," [Online]. Available: https://www.astmsteel.com/product/s50c-carbon-steel-jis-g4051/. [Accessed 2019].

"SSP 45° Spot & Chamfering Cutter," in Chain-Headway Catalog, Volume 6, 2019–2020, Taichung, Taiwan, pp. B669, 2019.

"Perthometer M1," in Mobile Roughness Measuring Devices , Mahr, pp. 5–12.

"ARCS," ARCS Precision Technology, [Online]. Available: https://www.arcs.tw/en/product.php. [Accessed 2020].

Smart Design, Science and Technology – Lam et al (eds)
© 2021 the Author(s), ISBN 978-1-032-01993-2

Application of augmented reality technology in factory equipment inspection: Taking Microsoft HoloLens 2 as an example

Wen-Yuh Jywe & Tung-Hsien Hsieh
Smart Machinery and Intelligent Manufacturing Research Center, National Formosa University, Hu-Wei, Yunlin, Taiwan

Liang-Yin Kuo & Yu-Han Lai
Department of Multimedia design & Institute of Digital Content and Creative Industries, National Formosa University, Hu-Wei, Yunlin, Taiwan

ABSTRACT: Modern science and technology are gradually advancing, and as factory instruments become more and more sophisticated, their maintenance is relatively difficult. When the equipment fails, the maintenance personnel may need to carry the paper manual to read and record in addition to the inspection and repair in person. The paper may be stained during the process, which is quite inconvenient.
Microsoft HoloLens 2 is a pair of smart-glasses with excellent mixed reality technology. It allows users to communicate with gestures or voice in augmented reality and provides development tools for companies to create interactive content.
In this study, the Unity engine is used to receive sensor data on equipment, and Vuforia is used to visualize the data. Finally, Microsoft HoloLens 2 is used to present the data to the end user. It aims to integrate equipment information through mixed reality technology to make it easier for maintenance personnel to perform maintenance, or the personnel without maintenance experience can also maintain the equipment in a simple way.

Keywords: Factory, Equipment maintenance, Augmented reality, Mixed reality, Microsoft HoloLens 2

1 INTRODUCTION

Nowadays, all kinds of machinery, electronics, automobiles and other industries need to manufacture metal parts through processing machines and other instruments. With the development of technology, instruments have become more and more sophisticated and with the maturity of robotics and AI technology, various fields have gradually realized fully automated factories, and related monitoring and management systems are a promising industry trend in the future.

MR (mixed reality) is a new technology rising in recent years. It is the combination of VR (virtual reality) and AR (augmented reality). This technology can receive the user's surrounding environment for recognition through the sensor on the device, and make the environmental data combine with the virtual objects of the system in the device. It enables users to see and interact with virtual objects in the physical environment, thereby enhancing the information that users can obtain in the environment. At present, the most representative device capable of implementing this technology is HoloLens 2 produced by Microsoft.

This study will combine AR technology and HoloLens 2 with the hope that the instrument status can be obtained more easily and quickly when personnel patrol the field.

2 EXPERIMENTAL

This study focuses on the convenience of obtaining information during field inspection as the main goal. To simulate the actual situation of the field, this study cooperated with National Formosa University Interdisciplinary Implementation Hall. There are 5-axis machine and 3-axis machine in the field to simulate the operation in the actual environment.

This study divides the experimental process into the following steps:

I. Install pressure, humidity and temperature sensors in the machine in the Interdisciplinary Implementation Hall.
II. Publish the obtained values to the server using the MQTT protocol.
III. Use Unity engine to get the data and draw it into easy-to-read graph information.
IV. Use the Vuforia in the Unity engine and the HoloLens 2 software development kit (SDK) Mixed Reality Toolkit (MRTK) to recognize and superimpose the image information in III. with the graphics card, and output to the HoloLens 2.
V. Have the users wear HoloLens 2 to watch the picture for experience.

DOI 10.1201/9781003188513-12

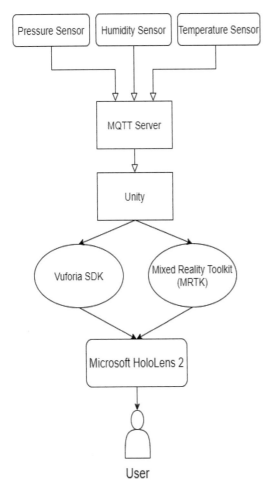

Figure 1.　Experimental Flow Chart.

3　RESULTS AND DISCUSSION

This section will explain the process and method of the experiment:

I. Install the sensor on the machine in the field.

II. Publish the obtained values to the server using the MQTT protocol.

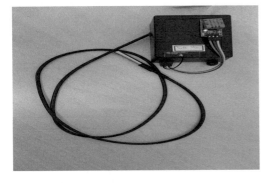

Figure 2.　Humidity and temperature sensors.

MQTT is a Client Server publish/subscribe messaging transport protocol. Because the minimum message size of MQTT is only 2 Bytes, it is suitable for IoT communications with limited bandwidth. MQTT protocol defines two network entities: message broker and client. Among them, the message broker is used to receive messages from clients and forward them to other target clients.

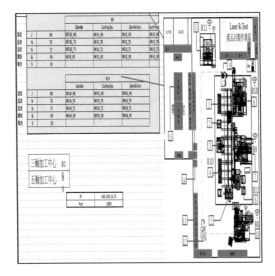

Figure 3.　Sensor distribution map in Interdisciplinary Implementation Hall.

Figure 4.　Processing machines in Interdisciplinary Implementation Hall.

III. Use Unity engine to get the data and draw it into easy-to-read graph information.

Write script in Unity to subscribe to the MQTT server, and then draw the obtained information into a graphical interface.

IV. Use the Vuforia in the Unity engine and the HoloLens 2 SDK MRTK to recognition and superimpose the image information in III. with the graphics card, and output to the HoloLens 2.

51

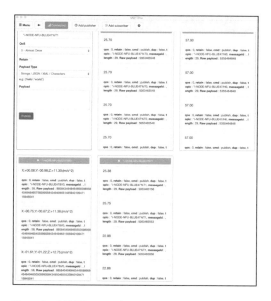

Figure 5. MQTT server data.

Figure 6. Unity script setting.

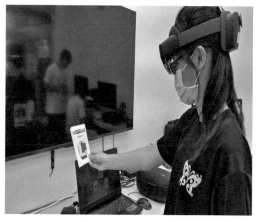

Figure 7. Image recognition using Vuforia and MRTK .

Vuforia is an augmented reality software development kit (SDK) for mobile and wearable devices that enables the creation of augmented reality applications. It uses computer vision technology to recognize and track planar images and 3D objects in real time.

Mixed Reality Toolkit(MRTK) is a Microsoft-driven project, used to accelerate cross-platform MR app development in Unity. It is can be used for HoloLens 2 and OpenVR devices (HTC Vive/Oculus Rift).

V. Users wear HoloLens 2 to watch the picture for experience.

Figure 8. Users wear HoloLens 2 to watch the picture.

Figure 9. Exhibition results at Interdisciplinary Implementation Hall.

4 CONCLUSION

At present, Microsoft hololens 2 has not been popularized in Taiwan. In addition to the high price and difficult acquisition, fewer manufacturers invest resources in the development of HoloLens 2 applications.

In recent years, the market trend is gradually approaching AR and VR. With the rapid development of mobile devices and portable devices, many industries are also developing related applications. Therefore, we hope to demonstrate HoloLens 2's MR application technology through this study, and let the industry know the potential and development of HoloLens 2.

REFERENCES

Shinji Chiba (2019). Microsoft HoloLens Technology and its Use. *The Imaging Society of Japan*, 58(3), 300–305.

Sebastian Langa, Mohammed Saif SheikhDastagir Kota, David Weigert, FabianBehrendt (2019) Mixed reality in production and logistics: Discussing the application potentials of Microsoft HoloLens. *Procedia Computer Science*, 149, 118–129.

MRTK documentation, Retrieved September 16, 2020 from https://microsoft.github.io/MixedRealityToolkit-Unity/README.html

Vuforia Augmented Reality SDK Wikipedia, Retrieved September 16, 2020 from https://en.wikipedia.org/wiki/Vuforia_Augmented_Reality_SDK

Shu-Yun Wei (2015) , *A Study and Development of Machine Tools Monitoring and Demonstration System*, Shih Chien University.

Michael Yuan (2020, January 7), Getting to know MQTT, Retrieved September 16, 2020 from https://developer.ibm.com/articles/iot-mqtt-why-good-for-iot/

Smart Design, Science and Technology – Lam et al (eds)
© 2021 the Author(s), ISBN 978-1-032-01993-2

Study of fast creating a 3D virtual character: A case of VRoid

Liang-Yin Kuo & Wei-Chen Wu
*Smart Machinery and Intelligent Manufacturing Research Center, National Formosa University, Yunlin, Taiwan;
Department of Multimedia design & Institute of Digital Content and Creative Industries, National Formosa
University, Hu-Wei, Yunlin, Taiwan*

ABSTRACT: With the development of digital technology, the word "VTuber" has become a popular word or concept since 2018. Due to the emergence of various virtual VTubers, in order to simplify the process for creators to virtual character modeling. This research uses the 3D character modeling tool "VRoid" developed by pixiv in Japan. Through VRoid, most of the hair, skeleton and material of the character can be completed in a short time. Skeleton and material, compared with the time spent on post-action correction of traditional modeling, save a lot of time for characters to make skeleton controllers. Compared with traditional 3D modeling tools such as 3d max, Maya and other software, the traditional 3D modeling tools binding skeleton process is cumbersome, and the concept of the skeleton controller is complicated, and it takes a lot of time resulting in the action not being smooth.

For this purpose, this research builds a virtual character from 2D graphic design, and then uses the VRoid to quickly build a character model from scratch. Using this character to apply the technology of a motion capture device, the virtual character in a fast modeling method will finally be used to demonstrate a variety of postures and actions. It provide a way for creators to improve efficiency and lower the threshold of 3D modeling when building character models.

Keywords: VRoid, Fast Modeling, Virtual Character, Skeletone Controller, Motion Capture

1 INTRODUCTION

Nowadays, with the development of virtual reality technology, virtual models can be used in a wide range, from computer games, movie effects and even the popular virtual anchor in recent years. Virtual YouTuber, also abbreviated as VTuber, which is often used to broadcast or upload movies on YouTube platform, has been deeply influenced by Japanese culture so far. The appearance of characters tends to Japanese animation style. The software VRoid is developed for pixiv in Japan, and the style of characters built in VRoid software is also Japanese animation style, which conforms to the character style to be designed in this study.

Virtual anchors are divided into 2D and 3D models. After some 2D creators use Live2D to create character models, they use webcams and FaceRig to realize the actions of the models. Since 2D models are limited to planes, the interaction between characters and other objects or audiences is less authentic, and 3D models can have more realistic interactive performance. However, compared with 2D planes, creators face more difficult technical problems in modeling. After using traditional time to build a character model, it is often encountered that the character's movements are not smooth, and afterwards, it takes a lot of time to adjust the movement of the skeleton.

This research will first create a virtual character appearance from 2D graphic design, and then use VRoid to build a 3D model and apply the character.

The technology of the dynamic capture device will eventually use a virtual character in a fast modeling method to demonstrate a variety of poses. The ultimate goal is to provide graphic creators with a way to improve efficiency and lower the threshold of 3D modeling when building character models.

2 EXPERIMENTAL

The visual design of virtual characters is very important. In the era of Internet media explosion, most people will decide whether to attract people to watch the content by their first impression and appearance. The design elements and color matching commonly used by virtual characters in the past are referred to when collecting information on appearance design. After the character appearance design is decided, the 3D model of the character is established by VRoid. The researchers hope that the virtual character designed

DOI 10.1201/9781003188513-13

can meet the preferences of all age groups, hope to use lively color matching and daily appearance, with smooth posture and action display, and hope to attract the public to continue to pay attention to this virtual role in the future.

This research will be divided into three parts: in the first part the 2D appearance of the virtual character will be designed. The theme of this role setting is "virtual anchor of National Formosa University". Based on the history of the University and the characteristics of the school, the shape of the virtual character will be extended and three views of the role will be drawn.

The second part is based on the three views of the role design in the first stage and explains how to use vroid to build the 3D model of the character and the display process.

The third part outputs 3D virtual characters and VRoid output files. VRM files support the development tool Unity. By placing the characters in Unity and applying the body movements recorded by dynamic capture devices to the character model, we can achieve rapid modeling and present dynamic and fluent virtual characters without modifying the skeleton.

3 RESULTS AND DISCUSSION

This section is divided into three stages, including character graphic design, VRoid modeling process, and applying dynamic capture. It will explain the creation process and experimental methods:

3.1 *Part 1: Character design*

At present, VRoid's built-in model only has two feet of human beings. Therefore, we design a Japanese animation style character. The background of the character is set as the virtual anchor of Taiwan Huwei University of science and technology. The creative idea is that the school started from the industrial department in the past, and the ratio of male to female in the school has become male. Therefore, the gender of the role is set as a female role with a neutral style as the idea and conception of character design. The personality of the characters is lively and cheerful, good at dancing, with strong contrast in color matching.

After the neutral and lively style of character design was established, the hairstyle was designed as light purple short hair with yellow hairpin. The clothing elements included a loose sports coat and canvas shoes, which were commonly used by college students. The arm design of the coat was the abbreviation of National Formosa University of science and technology. The main color matching of the characters uses the dark blue and yellow of the school emblem as a strong contrast, and the auxiliary colors are bright purple and pink. The overall selection of bright colors creates a lively atmosphere. It is expected that the 2D appearance designed in this study can make people feel the image of youth that represents the virtual anchor of the school.

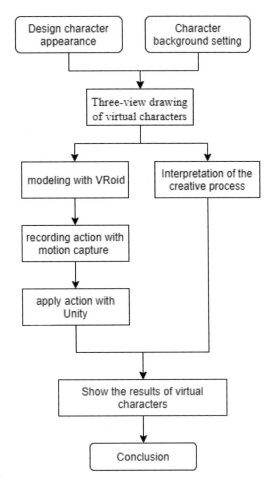

Figure 1. Process architecture.

Figure 2. Three-view drawing of virtual characters.

3.2 *Part 2: 3D modeling*

Using the characters designed in the first stage, there are three views as reference objects, which can effectively assist the process of 3D modeling.

I. First of all, we use the built-in avatar in vroid. The default character model in the software has been bound to the body skeleton. Here, you only need to adjust the details such as the proportion of the character's body and the position of the five facial features with the editor on the right.

Figure 3.   VRoid initial screen.

II. Pull out the appropriate mesh according to the shape of the design drawing, and pull out the hair along the grid. Hair details can be changed with different mesh layers.

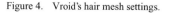

Figure 4.   Vroid's hair mesh settings.

III. Look for the built-in garment appearance model similar to the design as the basis for modification, and draw the texture of clothes in different parts of the character. VRoid clothing details can be adjusted less, although there is an editor to adjust the scale, but to add details, such as handheld objects, you need to combine with other 3D modeling tools.

Figure 5.   VRoid's clothing basic model and modified mark.

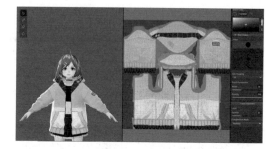

Figure 6.   VRoid clothes texture drawing.

IV. In VRoid, the built-in model action is provided for users to apply, and the action of the character can be viewed here without binding the skeleton.

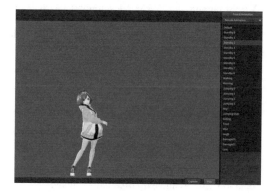

Figure 7.   Apply built-in actions.

V. After completion, the character model can be output. The "VRM file" output by vroid is the file format used to process the 3D avatar (3D model) data of VR application.

3.3   *Part 3: motion capture*

Put the output ".vrm" file into Unity, apply the action recorded by the motion capture device, and put it into the character model, so that can demonstrate a variety of poses and movements.

Figure 8.   Motion capture device.

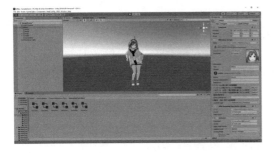

Figure 9. Virtual character roles applying actions in unity.

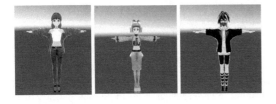

Figure 10. In the VTuber creation competition, the creator uses VRoid to model.

Using VRoid to achieve the purpose of rapid modeling, the virtual character can use the built-in skeleton system and only needs to fine tune the action to solve the problem of piercing the mold. VRoid is very convenient for the creator and can complete the 3D character model without spending a lot of time making the skeleton controller.

4 CONCLUSION

At present, there is more and more competition related to virtual characters and VTuber designs in Taiwan. For example, in the 2020 on campus VTuber creation competition of National Formosa University of science and technology, several groups of creators choose to use VRoid as a modeling tool.

Although VRoid modeling is fast in competition, its disadvantage is that there is no way to model anything other than bipedal humans, and there are many appearance limitations in built-in. It is still necessary to use 3D modeling tools such as Maya or 3D Max to make the appearance of characters closer to the original design. However, the skeleton system is quite convenient.

The field of virtual anchors will gradually become the trend of media, and the diversified content and interactivity may replace the traditional media. VRoid is a new modeling tool. Compared with other modeling tools, the network resources are insufficient. It is difficult for Chinese users to find information. I think VRoid is worthy of in-depth study. I hope this study can promote the convenience of VRoid skeleton binding and provide more creators to use to create 3D models of virtual characters.

REFERENCES

About Vroid. Retrieved September 15, 2020 from https://vroid.com/en/about
Yu-Hsuan Hung (2011). The Rig Control System of 3D Animation Software MAYA-the Study of Joint Control Mel.
VRM. Retrieved October 18, 2020 from https://vrm.dev/en/
Wikipedia-Virtual YouTuber. Retrieved September 15, 2020 from https://en.wikipedia.org/wiki/Virtual_YouTuber
Hui-Chun Wu (2007).A Study of Hair Design in Japanese Computer Romance Games Focusing on the Female Character

Smart Design, Science and Technology – Lam et al (eds)
© 2021 the Author(s), ISBN 978-1-032-01993-2

Discussion on how the choice of action figure's facial expression affects consumers' perception

Hsi-Jen Chen
Department of Industrial Design, National Cheng Kung University, Tainan, Taiwan

Chian-Fan Liou*
Department of Industrial Design, National Cheng Kung University, Tainan, Taiwan
Department of Visual Communication Design, Southern Taiwan University of Science and Technology, Tainan, Taiwan

ABSTRACT: Commodification of animated characters from the films can bring long lasting and amazing profits, be it the income from the copy license or the original release, which is far more than the box office earnings. The animated characters in the film usually have rich facial expressions. When an animated character is commodified into an action figure product, the designer needs to know if the choice of their facial expressions will affect consumers' perception and purchase motivations. This study is going to use the quantified questionnaire to do the survey and collect data of consumers' perceptions and purchase motivation towards the choice of action figure's facial expressions. The data will be analyzed through One-Way ANOVA, which can be used as a scientific reference for the future design teams or businesses to choose the right facial expression when they engage in turning an animated character to an action figure product. The research result summarizes the findings for how facial expressions affect consumer's preference and purchase motivations as below: In terms of consumer's preference, a laughing face is more appealing than a smiling or a sad face; a surprised face is more appealing than a sad face; a confused face doesn't make much of an impact compared with other facial expressions. In terms of consumer's purchase motivation, a laughing face is more appealing than a smiling one, a sad fase, and a confused fase, while a surprised face is more appealing than a sad face.

1 INTRODUCTION

According to the past published data, the box office earnings were never the source of main income for either animation movies or films as their branded products amount to 60% to 70% of the major revenues. For example, in 2004, Japan exported 46.2 billion Yen to the US market through the animation industry; whereas its authorized merchandise created 534 billion Yen value in the market via export, which was 10 times more than the film revenue itself (Li 2012). Undoubtedly, the branded products and royalty of copy license extended from animation films can create huge profits to the animation film and television industries.

The scholar Lee and Liao also indicated that the branded products don't just create a long-term profit but also raise the popularity of the original work in a lasting manner. For example, the Japanese animation "Neon Genesis Evangelion" still received revenues from its branded products and loyalties from its intellectual property even a decade after its film was out of theaters. The fashion of its diverse action figures

collection is well marketed, which leads to its growing popularity through the years (Lee & Liao 2010).

Figure was the word adopted by Hong Kong people for human figure puppets. In a narrow definition, it referred to an animated 12-inch sized human figure toys. As the culture evolves, it can now refer to any three-dimensional character toy or model in general. As the market evolves, an action figure is not only a simple character toy, it also comes with a narrative background story and a cultural setting. And yet this setting is not a functional design as it imposes more value to consumer's perception and sentiments than its practicality (Chang 2009).

Action figure products are marketed through different channels in Taiwan, such as toy model shops, convenience stores, instant food restaurants, claw machines, and capsule toys etc. as a gift or as a selling product. The popularity of action figures is not only limited to the collector market. It involves various commercial marketing and introduces the business speaker for the products, which creates very high commercial values (Norman 2005). The style of action figures mainly focuses on the character design, whether it is from comic books, animations, movies or the gaming industry. It brings

---

*Corresponding Author

DOI 10.1201/9781003188513-14

the main income for digital content industry and its relevant business. Those characters are not only marketed through their original background stories, but also applied into other accessory products and joined with other cross-industry business speakers (Huang & Li 2013).

The action figure created by the digital content industry has an existing high popularity thanks to media's report and promotion. It helps consumers remember and learn the knowledge of the products. Moreover, each character is already created with sharp features and background stories, thus it's easier for consumers to reflect their feelings on those characters and their stories and they become further attached to the product. The research found out that the successful shaping of a character in terms of their personality or looks will have people create sentimental attachments, which stimulates adults to spend money on the products. The appealing looks of the character will also attract children (Hung & Lin 2016). Therefore, in the process of character design, the designer usually creates a highly recognizable style for the character based on their features and personality. However, the styling and design of an action figure relies heavily on the designer's experience, knowledge and technique (Chen & Shen 2011). Ideally, it requires quantified research on consumer perception and purchase motivation for analysis for a scientific design principle to avoid a designer's subjective view of the product creating any risk in the market.

There was numerous research already done on the action figure design principles and animated characters design in the past, but mainly specific to their own area discussing how attraction works for their target customers. There was very little discussed on the transformation from an animated character to an action figure. Since animated characters usually contain their personalities and features created by the drama, they already come with various facial expressions and emotions. However, when they turn into action figures, they only have a certain choice for facial expression, which is usually decided by the designer based on their subjective opinions. To minimize the risk of misjudging the choice of facial expression during transformation, this study is going to discuss on how the application of facial expressions on action figures transformed from animated characters affect consumers' perceptions and purchase choices. Further we can conclude consumers' preference on an action figure's facial expressions as a scientific reference for future design of action figures transformed from animated characters.

To discuss consumer's preference of different facial expressions on action figures transformed from animated characters, we must:

1. Discuss if different facial expressions of action figures affect consumer's purchase motivation, and
2. Provide a reference for future designers and businesses who want to create action figure products transformed from animated characters.

2 EXPERIMENTAL

In terms of choice of animation film and action figure samples, to avoid tester's personal biased choice on certain animated characters, we skipped famous animated films. We selected a short animation film, "Dinner," which was released in 2010 and won a lot of awards in many competitions as a testing sample. The length of film is 9 minutes and 9 seconds with a resolution of 1920*1080, 72 dpi, a play speed of 24fps, with dual sound tracks and was played through a projector.

When applying the focus group method, we sought 6 specialists who design animated characters to agree on joining this research. Before the meeting started, we explained again the motivation and purpose of this research. We invited them to view the film to understand the plots and the character features. After that, we selected the character models. After discussion, we selected five presentative facial expressions, including: Smiling face, sad face, surprised face, laughing face and confused face.

Table 1. Specialist members.

Specialist	1	2	3	4	5	6
Gender	M	M	F	M	F	M
Seniority	12	10	6	6	5	5

After setting up 5 samples for questionnaires by applying the Focus Group method, we created flashcards for questionnaires. Five different facial expressions were chosen for the flashcard animated character for photo shooting. The background of the flashcard is plain white without any pattern. The camera was set in a fixed position for photographing with gentle lighting applied between the camera and the model. The animated character's posture and its styling outfit remained the same for the photo shooting for different facial expressions. After photo shooting, we used the computer program to remove the photo background from color to black and white. This was to focus on their facial expressions only and to minimize any possible factors that could affect the testers' perceptions. There were 5 photos chosen for their facial expressions, as follows:

Figure 1. Flashcards of animated character's different facial expressions (Smiling face, sad face, surprised face, laughing face, and confused face).

Due to the budget for this experiment, we mainly invited the students studying in relevant design subjects to join this test with a total number of 86 participants at the age between 18 and 22. As it was a higher number of people, we divided them into 2 groups for film watching. The space and audio setting of the watch room was exactly the same for the two audience groups. After finishing the film watching, they were given the online questionnaires to complete. In order to avoid the influence from word writing in the questionnaire, the five facial expressions in the flashcards were coded by A01 (smiling face), A02 (sad face), A03 (surprised face), A04 (laughing face), and A05 (confused face) in the flashcard size of 1108*1478 pixels, 72 dpi. Two questions were asked after the flashcards were shown.

1. I would love this animated character to be made into an action figure.
2. This facial expression attracts me and increases my motivation to buy this action figure.

The 7-level Likert scale was used to calculate the score for the answers. Scale 1 means "totally disagree" while scale 7 means "totally agree." From the calculation, we can analyze the different perceptions between the testers. This experiment collected 74 successful questionnaires. The data from the surveys were analyzed by One-Way Analysis of Variance (ANOVA) to discuss how the five different facial expressions of action figures affect consumer's preferences and their purchase motivations, which come from their perceptions.

3 RESULTS AND DISCUSSION

As One-Way ANOVA indicates, with the five facial expressions (smiling face, sad face, surprised face, laughing face and confused face) on an action figure transformed from an animated character in question 1: "I would love this animated character to be made into an action figure," there is not much difference in ANOVA, where $p = 0.305 > 0.05$ indicates homogeneity; the variance analysis shows there is an obvious difference between 5 samples: $F (4, 365) = 7.732$, $p < 0.01$. Scheffe post hoc tests indicate: A04 laughing face ($M = 4.8$, $SD = 1.535$) is obviously superior than A01 smiling face ($M = 4.04$, $SD = 1.389$) and A02 sad face ($M = 3.57$ $SD = 1.206$), whereas A03 surprised face ($M = 4.3$, $SD = 1.29$) is obviously superior than the A02 sad face ($M = 3.57$, $SD = 1.206$). However, A05 confused face ($M = 4.15$, $SD = 1.43$) hasn't shown much variance from other items.

In question No. 2: "this facial expression attracts me and increases my motivation to buy this action figure," there is not much difference in ANOVA, where $p = 0.435 > 0.05$ indicates homogeneity; the variance analysis shows there is an obvious difference between 5 samples: $F (4, 365) = 8.834$, $p < 0.01$. Scheffe post hoc tests indicate: A04 laughing face ($M = 4.61$, $SD = 1.629$) is obviously superior than A01 smiling face ($M = 3.74$, $SD = 1.434$), A02 sad face ($M = 3.26$, $SD = 1.385$) and A05 surprised face ($M = 3.84$, $SD = 1.405$), whereas A03 surprised face

($M = 4.16$, $SD = 1.405$) is superior than A02 sad face ($M = 3.26$, $SD = 1.385$).

4 CONCLUSION

From the questionnaire statistic result, we learn that when a product is made from an animated character to an action figure, the choice of facial expressions will much affect consumer's perceptions. Therefore, the product designer must select and think carefully on the choice. Even if the same animated character design will produce different perceptions on consumers. The variance between each item is as follows:

1. When the laughing face is applied to an action figure, consumer's purchase motivation and preference are both stronger than when the smiling face and the sad face are applied.
2. In terms of purchase motivation, the laughing face is more popular than the confused face; in terms of what attracts consumer's attentions, the laughing face is not much different from the confused face.
3. Whether in purchase motivation or preference, the surprised face is more popular than the sad face.
4. There is no obvious variance between the laughing face and the surprised face.

According to the current research result, between facial expression applications, although there is no obvious variance between the laughing face and the surprised face, the laughing face is still a superior choice when it comes to designing an action figure transformed from an animated figure than the other facial expressions as a conservative decision. From the other side, the sad face received negative feedbacks when compared to both the laughing face and the surprised face. In the future, we advise that the designer should choose the laughing face for the action figure's facial expression to attract consumers, followed by the surprised face, and the least for the sad face, which is the least to most people's liking. Unless necessary, the sad face should be avoided as a facial expression.

The suggestions for future researches are as follows: Due to the limited budget, the surveys were only carried on to a group of undergraduate students. Although the students studying in design subjects have better sensitivity in preference and purchase motivation for action figures, still the data is not sufficient to speak for all the consumers. In choosing the animated characters, we suggest the choices for different types of animated films and commodified characters should be increased. Moreover, a qualitative interview for cross analysis is also encouraged for further and more robust data analyses to study how an animated character can be transformed to an action figure product successfully in the market.

REFERENCES

Chang, E. E., 2009, "Deconstruction of the "Figure" Socio-Cultural Phenomenon", Taiwan: A Radical Quarterly in Social Studies, Vol. 73, pp. 167–188.

Chen, C. C. and Shen, Z. X., 2011, "The Study of Consumer Preference for Traditional Culture Creativity Design-A Case of Figure Designs", Journal of National Taiwan College of Arts, Vol. 89, pp. 127–150.

Huang, K. L. and Li, C. H., 2013, "A Study of the Tainan Cultural Based on the New Generations' Perspectives on Action Figure Design", Journal of Cultural and Creative Industries Research, Vol. 3 No. 4, pp. 147–155.

Hung, P. H. and Lin, R. T., 2016, "A Study of the Correlation between Color Image and Modeling Design on Virtual Characters' Personality", Journal of Ergonomic Study, Vol. 18 No. 1, pp. 45–56.

Lee, C. M. and Liao, H. W., 2010, "The Study of Key Successful Factors of Taiwanese Animation Movies to Attract Audience by AHP", Management Research, Vol. 10, pp. 1–35.

Li, S. H., 2012, "Riben meiti neirong chanye zhi wandai lianmeng moshi: xiankuang yu tiaozhan (in Chinese)", Industry Management Review, Vol. 5 No. 2, pp. 39-54.

Norman, D. A., 2005, "Emotional Design: Why We Love (or Hate) Everyday Things". New York: Basic Books.

Smart Design, Science and Technology – Lam et al (eds)
© 2021 the Author(s), ISBN 978-1-032-01993-2

An investigation of problematic smartphone use by Taiwan's children and young people (2-12 years old)

Siu-Tsen Shen*
Department of Multimedia Design, National Formosa University, Taiwan

Stephen D. Prior
Faculty of Engineering and Physical Sciences, The University of Southampton, UK

ABSTRACT: The rise of smartphones and the role of Social Media (Facebook, Twitter, WhatsApp, YouTube, etc.) have brought about concerns over the mental health, well-being and happiness of the Children and Young People, CYP (2-12 years old) population. The rise in the use of so-called smart devices, with some people reportedly using these for up to 8 hours per day, raises concerns about whether these devices are liberating or inhibiting people's productivity and mental & physical health. The World Happiness Report (2018) surveyed roughly 3,000 respondents from 156 countries, asking them to evaluate their current lives on a scale where 0 represents the worst possible life and 10, the best possible. For the world as a whole, the distribution is normally distributed about the median answer of 5, with the population-weighted mean being 5.264.

1 INTRODUCTION

1.1 *The use of smartphones and their impact on children and young people*

Due to the rapid growth of smartphone use, children and young people (CYP) are spending more and more time in the digital world and are becoming addicted. A recent review paper from the UK identified 924 studies of smartphone use amongst children and young people. They concluded that approximately 1 in 4 CYP are exhibiting Problematic Smartphone Use (PSU) and could therefore be described as addicted to their smartphone, with all the negative mental health implications that this implies (Sohn et al. 2019). Research into the use of smartphones by the very young (<12 years old) is largely unreported and is an area of vital importance.

The average adult uses nine apps on a daily basis, and 30 apps per month and spends 3.1 hours a day using apps (Nick 2019). Statistics shows that 100% of 18 to 29 years old in the USA access the internet regularly. The youngest age group (18-24) uses the largest number of app hours (3.2).

Figure 1 shows that the age range of 0 to 11 years has a penetration rate of 11.1% using smartphones (eMarketer 2019). However, this may well be under-reported. The authors hypothesize that busy parents may be using the ubiquitous smartphone to pacify their very young children at times of high stress, such as mealtimes, bedtimes and during outside working

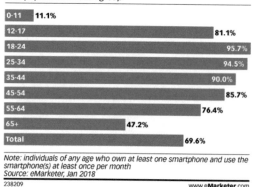

US Smartphone User Penetration, by Age, 2018
% of population in each group

Age	
0-11	11.1%
12-17	81.1%
18-24	95.7%
25-34	94.5%
35-44	90.0%
45-54	85.7%
55-64	76.4%
65+	47.2%
Total	69.6%

Note: individuals of any age who own at least one smartphone and use the smartphone(s) at least once per month
Source: eMarketer, Jan 2018
238209   www.eMarketer.com

Figure 1.   US smartphone user penetration by age 2018 (eMarketer 2019).

hours. Depending on your perspective, this could be storing up trouble for the future.

According to a recent Pew Research survey, 1 in 3 American adults say they are online almost constantly, with another 45% claiming to go online several times per day.

Roughly half of 18- to 29-year-olds (48%) say they go online almost constantly and 46% go online multiple times per day. By comparison, just 7% of those 65 and older go online almost constantly (Andrew & Madhu 2019).

Teenagers seem to find it even harder to disconnect, with nearly half of U.S. teens aged 13 to 17 saying they

---

DOI 10.1201/9781003188513-15

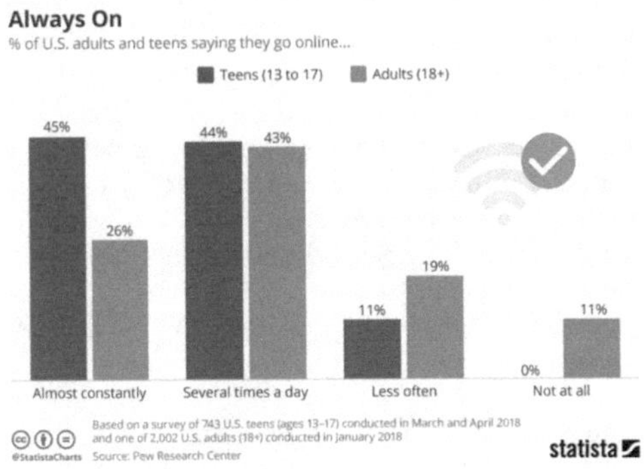

Always On
% of U.S. adults and teens saying they go online...

■ Teens (13 to 17) ■ Adults (18+)

45% | 26% | 44% | 43% | 11% | 19% | 0% | 11%
Almost constantly | Several times a day | Less often | Not at all

Based on a survey of 743 U.S. teens (ages 13–17) conducted in March and April 2018
and one of 2,002 U.S. adults (18+) conducted in January 2018
@StatistaCharts Source: Pew Research Center

statista

Figure 2. U.S. teens online pattern (Felix 2018).

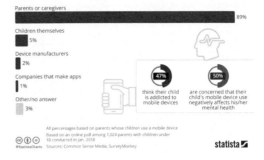

U.S. parents' view on who's primarily responsible for limiting children's mobile device use

Parents or caregivers 89%
Children themselves 5%
Device manufacturers 2%
Companies that make apps 1%
Other/no answer 3%

47% think their child is addicted to mobile devices
50% are concerned that their child's mobile device use negatively affects his/her mental health

All percentages based on parents whose children use a mobile device
Based on an online poll among 1,024 parents with children under 18 conducted in Jan. 2018
@StatistaCharts Sources: Common Sense Media, SurveyMonkey

statista

Figure 3. Who's responsible for children/s smartphone use (Felix 2018).

are constantly online (see Figure 2). The main reason for this trend is the ubiquity of smartphones which 95% of teenagers say they have access to. Compared to a similar survey conducted four years ago, the share of teens who are always online has nearly doubled, suggesting that young people's digital lives demand more and more attention and threaten to crowd out real-life personal interactions (Felix 2018).

A wide range of scientific studies have showed that the use of smartphones in bed does affect quality of sleep. According to Common Sense Media, 40% of teenagers and 26% of parents in the U.S. check their smartphone within five minutes of going to bed, and a significant proportion of Americans even use their mobile device in the middle of the night.

Even though smartphones are equipped with a night mode to limit the negative effects of bedside smartphone use, these are merely filtering out blue light which suppresses the secretion of melatonin, a hormone that regulates sleep cycles. What these "night modes" cannot do is limit the negative effects that smartphones and their steady stream of stimulation have on people's ability to unwind and fall asleep (Felix 2019).

It is said that the use of smartphones is not a good habit and not healthy in bed. However, we still cannot help but bring it to our bedside. Interestingly, research has shown that early morning smartphone use is nearly as prevalent as it is late at night. Instead of waking up slowly, brushing our teeth, and sorting our thoughts for the day ahead, people tend to grab their smartphone, and check their messages or emails straight away.

Tech addiction is a growing concern in modern societies, as more and more people realize that they may be spending too much time with their phone instead of engaging in real world activities. Several studies have linked smartphone use with negative effects on mental and physical wellbeing with children, seen as particularly vulnerable to the distractive lure of the small screen.

Common Sense Media found that many parents are in fact concerned with the smartphone use of their children, with nearly half of them saying their child might be addicted to his or her mobile device. Most parents

see themselves as primarily responsible for limiting their children's device use, but having more advanced parental controls would certainly help them to do so (Felix 2018).

Nowadays, smartphones and other digital devices enable children to get connected from a very young age. Ninety-eight percent of kids under 8 years old have access to a device at home, and 50% of teenagers say they feel addicted to their phones (Sense 2019). The impact of these dramatic increases will be felt most amongst the younger generation. In order to protect vulnerable children, some schools have now banned smartphones during the school day, as well as asking parents to prohibit their use during periods of the weekend (Hymas 2018).

1.2 *Nomophobia*

Several studies into smartphone and internet addiction have proved that this level of use is harmful to our minds. Researchers have even coined a new term "Nomophobia" to describe a person who has the fear of not being able to use their smartphone or other digital device. Yildirim and Correia (2015) investigated Nomophobia as a theoretical construct among young adults, and dimensions of Nomophobia were identified and described. The Nomophobia Questionnaire (NMP-Q) was devised and validated with 301 undergraduate students. The NMP-Q can be used as a measure of Nomophobia and can be considered a situational phobia and is included in the Diagnostic and Statistical Manual of Mental Disorders (DSM-5) by the American Psychiatric Association (Yildirim & Correia 2015).

According to CNN, new research has shown that being addicted to our smartphones can affect our safety and our health. Furthermore, new research also reveals a connection to a lack of neural chemicals that affect people's ability to focus (LaMotte 2017). Moreover, Common Sense Media found that 50% of teens feel that they are addicted to their mobile devices, and nearly 80% of them check their phones hourly; 72% feel the need to respond immediately (Wallace 2016).

Several Apps have recently emerged to help people to monitor and control their addiction to their smartphones, such as Moment and Mute which are trackers to analyse the time you spend on your phone, and let you set limits for different apps or times.

2 THE RESEARCH STUDY

2.1 Research questions

The research questions in this study involve issues of parental decision making of smartphone use amongst their children.

(A) How is Taiwanese parental control exercised with regards to their children's smartphone use?
(B) What is the motivation for child smartphone users to engage with a social media communication application on their devices?
(C) To what degree are children able to resist the temptation to interact with their smartphone?
(D) What is the context of daily use of child (2-12 years) smartphone users?

In order to better understand the research problem and to help answer the research questions, a controlled study was carried out with the help of 63 parent volunteers in the areas shown below (the research question they relate to can be seen in brackets):

1. Investigate happiness scales from different types of young smartphone users (A).
2. A study into young user knowledge and contextual influences (B & C).
3. A review of work done on young smartphone usage issues related to social communication app category (B & C).
4. A further user study comparing and contrasting different approaches to young user behavior and preference based on other influential factors (D).

2.2 Hypotheses

This study investigated the level of happiness of Taiwanese parents, together with their children's use of smartphone devices.

- The study hypothesis (H1) states that busy Taiwanese parents' reliance on their own smartphone devices has had a negative effect on their parenting role.
- The study hypothesis (H2) states that Taiwanese children and young people (CYP) are becoming addicted to their smartphones.
- The study hypothesis (H3) is that CYP will resist the temptation to disengage with their smartphone devices.

2.3 Research plan

Increasingly, happiness is the proper measure of social progress and the goal of public policy. The rising use of so-called smart devices, with people reporting using these for up to eight hours per day on average,

raises concerns whether these devices are liberating or inhibiting people's productivity and children's well-being and mental health.

This study investigated the level of happiness of Taiwanese parents, children and young people with their use of smartphone devices. The user trial consisted of 63 parents with at least one child. After interviewing the potential participants, the selected participants who agreed to share personal information on their smartphones for this research were then asked various questions about their smartphone use. This research conducted a series of three online questionnaires, interviews, to observe users' cognitive behavior with smartphones over a one-year timeframe.

3 SURVEY RESULTS

3.1 Findings

The initial battery of survey questions provided a good insight into the general characteristics of the 63 survey participants and their 125 children (see Figure 4). Most were fairly young parents (20–35), 46% had one child, 44% had two children, 6% had three and 3% had four or more. In terms of gender, the first child born was biased towards being male (40) to female (22). The parents had a number of years of smartphone use. The participants gender was two-thirds female and one-third male. The vast majority being in full-time employment, some were studying for master's qualifications. Roughly 30% were using their phones between 5–6 hrs/day.

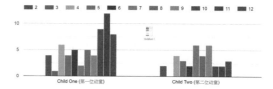

Figure 4. Children's age distribution – Child 1 & 2 (Years).

Almost 51% of the children had their own smartphone or tablet, 78% of these were recycled. 92% of the children were not allowed to take their smartphone to bed at night. Of the 8% that were, almost 15% of them reported poor sleep quality. The most popular Apps were YouTube, Facebook and Line.

In terms of negative effects of smartphone use, parents mentioned bad posture, poor educational attainment, sleeping disorders and anxiety.

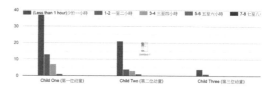

Figure 5. Children's daily smartphone use (hr/day).

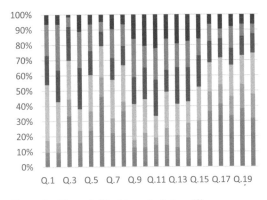

Figure 6. Nomophobia data analysis (n = 63).

In terms of the online smartphone happiness index, the vast majority of parents (76%) stated that it was a bad idea to let their children watch/use their smartphones/tablets. 81% believed that the use of a smartphone/tablet brought about negative effects. Most parents worried about damaged eyesight, followed by bad posture, lack of attention, poor performance at school and poor sleep.

When asked if they worried that their child would become addicted to their smartphone/tablet, 75% agreed. In a follow-up question, when asked if they thought their child was already addicted, 62% agreed.

This clearly shows an awareness of the problem, even though few participants were prepared to stop their children using this ubiquitous and (perceived as being) vital device.

As to why they would allow use of a smartphone/tablet, there was a wide range of answers, from when parents were busy, in an emergency, at eating times and at bedtime. In terms of smartphone use outside of the home environment, the most popular reason given was when visiting restaurants, followed closely by social events and then long journeys (car, bus, train, airplane).

Over 90% of the parents sampled had to tried to stop their children using smartphone devices. Of these, 54% described an angry response, 12% very angry response, and only 2% a happy response. Of the 66% of parents that observed a negative reaction, only 10% stated that they would give up trying this again.

In terms of overall parental happiness with their children's life, an overwhelming majority were happy, with a peak score of 8/10.

3.2 Nomophobia results

All of the participants were surveyed using the Nomophobia questionnaire (20 questions) as discussed in Section 1.2 above (see Figure 6).

In terms of the Nomophobia analysis. Of the 63 participants, the average score was calculated as being 81.7, meaning that they were classified as having a moderate Nomophobia level (see Table 1). This is 10% higher than an earlier study with young Taiwanese adults (Shen 2019).

Table 1. Nomophobia score and level.

Score	Nomophobia Level
NMP-Q Score = 20	Absent
21 ≤ NMP-Q Score < 60	Mild
60 ≤ NMP-Q Score < 100	Moderate
100 ≤ NMP-Q Score ≤ 140	Severe

4 CONCLUSION

This study has highlighted the problems posed by modern technology, and the impact that this has had on parents and their children. These individuals (borne to the internet generation), have known nothing else but the instant access and gratification that can be had by being connected in every sense to the society around them. However, this has been proved to also have a negative effect on health, wellbeing and communication skills. Parents are clearly aware of the issues and are trying to manage this. However, they lack clear guidance and advice. How we educate, inform and manage this situation is up to the individual user. However, as responsible parents, we owe it to our offspring to be both sensitive to their needs, and mindful of the consequences of too much screen time.

REFERENCES

Andrew, P. and K. Madhu. (2019). "About three-in-ten U.S. adults say they are 'almost constantly' online." Retrieved December19, 2019, from https://www.pewresearch.org/fact-tank/2019/07/25/americans-going-online-almost-constantly/.

eMarketer. (2019). "US Smartphone User Penetration, by Age, 2018 (% of population in each group)." Retrieved December 18, 2019, from https://www.emarketer.com/chart/219283/us-smartphone-user-penetration-by-age-2018-of-population-each-group.

Felix, R. (2018a). "Always On." Retrieved December 19, 2019, from https://www.statista.com/chart/14088/frequency-of-internet-usage-in-the-united-states/.

Felix, R. (2018b). "Who's Responsible for Children's Smartphone Use?" SMARTPHONE ADDICTION Retrieved December 19, 2019, from https://www.statista.com/chart/13052/smartphone-addiction-among-children/.

Felix, R. (2019). "America's Favorite Bedside Companion?" SMARTPHONE USAGE Retrieved December 19, 2019, from https://www.statista.com/chart/12017/smartphone-use-in-the-morning-and- at-night/.

Hymas, C. (2018). Secondary schools are introducing strict new bans on mobile phones. The Telegraph. London, UK, Telegraph Media Group Limited.

LaMotte, S. (2017). "Smartphone addiction could be changing your brain." Retrieved December 19, 2019, from https://edition.cnn.com/2017/11/30/health/smartphone-addiction-study/index.html.

Nick, G. (2019). "51 Jaw Dropping App Usage Statistics & Trends 2019." Retrieved December 18, 2019, from https://techjury.net/stats-about/app-usage/#gref.

Sense, C. (2019). "Digital Well-Being Is Common Sense." Retrieved December 19, 2019, from https://www.commonsensemedia.org/digital-well-being#get-resources.

Sohn, S., P. Rees, B. Wildridge, N. J. Kalk and B. Carter (2019). "Prevalence of problematic smartphone usage and associated mental health outcomes amongst children and young people: a systematic review, meta-analysis and GRADE of the evidence." BMC Psychiatry 19(1): 356.

Wallace, K. (2016). "Half of teens think they're addicted to their smartphones." Retrieved December 19, 2019, from https://edition.cnn.com/2016/05/03/health/teens-cell-phone-addiction-parents/index.html.

Yildirim, C. and A.-P. Correia (2015). "Exploring the dimensions of nomophobia: Development and validation of a self-reported questionnaire." 49: 130-137.

Smart Design, Science and Technology – Lam et al (eds)
© 2021 the Author(s), ISBN 978-1-032-01993-2

A study on the appropriate lighting level for hospital waiting rooms

Tai-Ming Huang*
Ph.D Program of Mechanical and Energy Engineering, Kun Shan University, Tainan, Taiwan

Huann-Ming Chou
Department of Mechanical Engineering, Kun Shan University, Tainan, Taiwan

ABSTRACT: For patients in Taiwan, the low medical cost and great convenience has resulted in overly lengthy waiting time. Some studies have shown that overly lengthy waiting time is the main reason contributing to patients' dissatisfaction towards their experience of medical treatment received. By improving the waiting area environment, patients' anxiety while waiting can be improved, thereby enhancing their satisfaction towards medical treatment received. In this study, the lighting environment of the hospital's waiting area was targeted. In order to find the most appropriate lighting brightness, the brightness of the waiting area was changed to three different shades, and a questionnaire survey was administered on the patients. The survey results show that adjusting the brightness of the waiting area lighting indeed affected the subjects' perceptive judgement of the visual space. When the light was too dim (250 LUX), the subjects' nervousness and sense of insecurity increased. However, when the light was too bright (430 LUX), it did not make the subjects more relaxed, more secure, nor more satisfied. Optimum lighting (370 LUX) catered to both the need for comfort and the need to save energy.

Keywords: Waiting Area, Lighting Brightness, Satisfaction.

1   INTRODUCTION

In Taiwan, the implementation of National Health Insurance has resulted in low cost and high convenience medical treatment for patients, bringing along an astonishing number of medical consultations. The huge number of people seeking medical treatment has of course caused the waiting time to be lengthened and the time that patients need to spend in the waiting room has also increased. Studies have shown that a significant number of patients are dissatisfied with the length of waiting time, which is the least satisfying aspect among the patient satisfaction (Kao 2000).

A pleasant waiting room with a comfortably, reasonably, and efficiently lit hospital environment is crucial in relieving patient pressure while waiting for a medical consultation. This study aimed to investigate the impact of the waiting room lighting levels on the patients' psychological feelings and to determine the appropriate lighting conditions for Taiwanese patients. Therefore, the research took place at a regional hospital in central Taiwan to examine the impact of the different lighting levels in the waiting room on the psychological feelings of patients who were waiting for a medical

consultation, so as to find the appropriate lighting conditions to relieve their pressure wherein, as well as to provide a reference for the design of future lighting environments in regard to hospital waiting rooms.

2   LITERATURE REVIEW

Waiting rooms are the area in which patients stay and wait before entering a consultation room. Among the relevant studies on outpatient satisfaction, the service attitude is the most important and the most satisfying aspect for outpatients, while the long waiting time being the most dissatisfying aspect (Chang et al. 2007).

In the study of "The Psychological Benefits of Waiting Room of Hospital," Hsieh Ya-ting found that different types of greening in the waiting rooms had partial effects on users' spatial perception, spatial cognition and psychological benefits; and a green environment indeed brought positive spatial perception, spatial cognition and psychological benefits to the users. She also believed that the green waiting room environment could improve the user's color perception, thermoception and facility comfort perception and thus further reduce the user's sense of heaviness (Hsieh 2007).

*Corresponding Author

DOI 10.1201/9781003188513-16

3 RESEARCH DESIGN AND METHODOLOGY

3.1 Research framework

By collecting the data from and reviewing the previous studies, with reference to the theories of the environmental psychology and buildings' indoor environment, this study was based on the factor that the waiting room environment could affect the psychological feelings of the users. Through literature review, the important environmental factors that had an impact on users' psychological feelings were organized, and the lighting conditions among which were studied (Chen 2002; Francis 1995; Xu & Yang 2005).

The research was conducted at a regional hospital in central Taiwan, Feng Yuan Hospital by the Ministry of Health and Welfare. Meanwhile, a questionnaire survey was employed on the users of the waiting room at the Department of General Internal Medicine under different lighting conditions, in order to understand their psychological feelings and satisfaction wherein.

In this experiment, we compared the impacts of the three different lighting background by changing the lighting levels in the same waiting room:

(1) The Brightest Group: ceiling recessed down light fixtures at the waiting room, four lit tubes, lighting levels of 330-570 LUX, with an average brightness of 450 LUX.
(2) The Middle Bright Group: ceiling recessed down light fixtures at the waiting room, reduced to three lit tubes, lighting levels of 270-490 LUX, with an average brightness of 370 LUX.
(3) The Darkest Group: ceiling recessed down light fixtures at the waiting room, reduced to two lit tubes, lighting levels of 130-300 LUX, with an average brightness of 230 LUX.

3.2 Questionnaire design

The questionnaire (as shown in Table 1) consists of 10 questions on a 5-point Likert Scale, with the scale scores divided into 5 levels ranging from strongly agree to strongly disagree.

3.3 Experimental period and procedure

The experimental period was from January 24 to March 2, 2018. The Middle Bright Group was the first group to be tested, with the ceiling recessed down light fixtures of the waiting room at the Department of General Internal Medicine as the experimental model; meanwhile three tubes were lit. The research subjects completed the questionnaire according to their perceptions with this lighting environment, yet the questionnaires with incomplete responses were eliminated until fifty questionnaires with complete responses were collected.

Then, the lighting fixtures were reduced to two lit tubes for the Darkest Group experiment, from where another fifty questionnaires with complete responses were collected in the same method during the same

Table 1. The questions and concepts of the impact of waiting room lighting levels on the patients' psychological feelings and satisfaction.

Questions	Patient's Psychological Feelings Evaluation
1 I feel relaxed or anxious in this waiting room.	
2 I would feel scared or safe in this waiting room.	
3 I feel a lot of pressure in this waiting room.	
4 Patients waiting for a medical consultation can be well taken care of here.	

Questions	Patient Satisfaction Evaluation
5 I can feel the hospital's dedication to taking care of the patients who are waiting for a medical consultation.	
6 I can do whatever I want in this waiting room.	
7 I'm restricted in this waiting room.	
8 This waiting room is perfect for my needs.	
9 I'm satisfied with this waiting room.	
10 I'm satisfied with this hospital.	

clinic hours. Finally, all four tubes of the ceiling recessed down light fixtures were lit for the Brightest Group experiment, and the last fifty questionnaires with the complete responses were collected during the same clinic hours.

4 RESEARCH FINDINGS

4.1 Basic information of the research subjects

There were 150 subjects in this study, of which 84 were males and 66 were females, accounting for 66% and 44% of the total subjects respectively, who aged between 10 to 80 years old.

4.2 Questionnaire results

4.2.1 Analysis on the impact of the lighting levels on the patients' psychological feelings

In the questionnaire design, questions No. 1–4 were used to test the impact of different lighting levels on the three groups of the respondents' psychological feelings. Except for question No. 3, which is a negative question The higher the score of the other questions, it means the more relaxed or secure the respondents are with the waiting room and hospital care. (as shown in Table 2)

Table 2.　The mean value of the impact of lighting levels on the patients' psychological feelings.

No.	Patients' Psychological Feelings Evaluation	The Mean Value Brightest Group	The Mean Value of the of the Middle Bright Group	The Mean Value of the Darkest Group
1	I feel relaxed or anxious	2.3	2.7	2.3
2	I would feel scared or safe	3.4	3.6	2.9
3	I feel a lot of pressure	2.9	2.9	2.8
4	Patients waiting for a medical consultation can be well taken care of here.	3.0	3.3	2.9

Table 3.　The ratings of the impact of lighting levels on the patients' psychological feelings.

Level	Diff	lwr	upr	p
2-1	.21	−.05	.46	.14
3-1	−.18	−.43	.08	.24
3-2	−.28	−.63	−.12	<.01**

Level 1: The Brightest Group, 2: The Middle Bright Group, 3: The Darkest Group

The multiple comparison method of Tukey's HSD Post-Hoc Test was used to statistically compare the differences, the results were shown in Table 3.

Based on the above statistics, the following conclusions can be drawn: If the brightness of the waiting room is too dark, the more nervous and insecure the respondents will feel. However, increasing the lighting level of the waiting room to 270-490 LUX, with an average brightness of 370 LUX, can make the respondents feel more relaxed and safer psychologically. Comparing it to the Brightest Group, the brighter environment will not make the respondents feel more

relaxed psychologically, instead it may increase their psychological pressure and sense of insecurity.

4.2.2　Analysis on the impact of the lighting levels on patient satisfaction

In the questionnaire design, questions No. 5-10 were used to test the impact of different lighting levels on the three groups of the respondents' satisfaction. Except for question No. 7, which is a negative question, the higher the score of the other questions, it means the more satisfied the respondents are with the waiting room and hospital care. (as shown in Table 4)

The multiple comparison method of Tukey's HSD Post-Hoc Test was used to statistically compare the differences between these three groups the results were shown in Table 5.

In terms of evaluating the impact of lighting levels on patient satisfaction, the group with the darkest lighting experienced the lowest satisfaction, while the group with the middle brightness experienced the highest satisfaction, indicating the brighter of the waiting room not necessarily the higher satisfaction of the respondents.

Table 4.　The mean value of the impact of lighting levels on patient satisfaction.

No.	Patient Satisfaction Evaluation	The Mean Value of the Brightest Group	The Mean Value of the Middle Bright Group	The Mean Value of the Darkest Group
5	I can feel the hospital's dedication to taking care of the patients who are waiting for a medical consultation	3.18	3.24	2.98
6	I can do whatever I want	2.68	2.74	2.52
7	I'm restricted.	2.92	3.08	2.72
8	My needs are fully met.	2.48	2.86	2.52
9	I'm satisfied with this waiting room.	2.78	3.04	2.52
10	I'm satisfied with this hospital	3.08	3.22	2.96

Table 5. The ratings of the impact of lighting levels on the patients' psychological feelings.

Level	Diff	lwr	upr	p
2-1	.18	−.08	.44	.21
3-1	−.15	−.41	.11	.36
3-2	−.33	−.59	−.07	<.01**

Level 1: The Brightest Group, 2: The Middle Bright Group, 3: The Darkest Group

5 CONCLUSION

In order to provide a better waiting room environment, particularly the lighting conditions, this study aimed to determine the lighting level that is both comfortable and energy-efficient for waiting rooms. The experiment of changing three different lighting levels in a same waiting room was conducted to investigate the impact of lighting levels on the respondents' spatial perception, psychological feelings, and patient satisfaction.

The study compared three lighting levels in three groups and found that the Darkest Group (the lighting level of 130-300 LUX, with an average brightness of 230 LUX) experienced the highest psychological stress, and the sense of insecurity, with the lowest patient satisfaction. The Middle Bright Group (the lighting level of 270-490 LUX, with an average brightness of 370 LUX) experienced the highest psychological relaxation and patient satisfaction, indicating that the dark background lighting of the waiting room was not conducive to the medical consultation. However, the psychological feelings and patient satisfaction of the Brightest Group (the lighting level of 330-570 LUX, with an average of 450 LUX) did not improve and might have negative effects instead. Hence, the experiment results suggest that the lighting level of 270-490 LUX, with an average brightness of 370 LUX, is more appropriate for waiting rooms. Moreover, an overly bright lighting background also means more energy consumption, which is not conducive to green-building concepts. Thus, it is not recommended that the average brightness of the waiting room to be as high as 450 LUX, but consider lowering it below 430 LUX.

REFERENCES

C.S. Chang, H.C. Weng, T.H. Hsu, 2007. Exploring the Medical Service Quality from the Perspective of Service Encounter-Example of the General Medicine and Gynaecology, Journal of Quality, 14(3), pp. 301–315.

M.S. Chen, 2002. Universal Design.

T.M.A. Francis, 1995. Environmental Psychology, Taipei: Wu-nan Book Inc.

Y.T. Hsieh, 2007. The Psychological Benefits of Waiting Room of Hospital (Master's thesis, Graduate Institute of Horticulture, National Taiwan University, Taipei, Taiwan).

M.Y. Kao, 2000. The Satisfaction of Medical Services on Hospital Among Patients with Chronic Illness, Taiwan Journal of Family Medicine, 10(4), pp. 212–225.

L.Q. Xu, G.X. Yang, 2005. Environmental Psychology: Environment, Perception and Behavior, Taipei: Wu-nan Book Inc.

Smart Design, Science and Technology – Lam et al (eds)
© 2021 the Author(s), ISBN 978-1-032-01993-2

A study on the effects of prenatal music education in reducing expectant mothers' anxiety during prenatal visits

Hung-Chuan Yu*
Ph.D Program of Mechanical and Energy Engineering, Kun Shan University, Tainan, Taiwan

Huann-Ming Chou
Department of Mechanical Engineering, Kun Shan University, Tainan, Taiwan

ABSTRACT: The purpose of this study is to explore whether the application of prenatal education music can reduce anxious emotions experienced by pregnant women. The subjective anxiety scale, and an outpatient clinic satisfaction survey, coupled with objective physiological responses and placental function monitoring results were used to carry out data collection and compilation. The situation of receiving medical treatment was also statistically analyzed. In this study, data were collected through the outpatient clinics of a hospital. The data collection time period was from March 1, 2017 to December 31, 2017, a total of 10 months. The cases were accepted at an obstetrics and gynecology hospital in Central Taiwan. During the case acceptance period, the pregnant women who had undergone ten complete prenatal checkups were included as research participants. A total of 100 valid samples were completed. Among them, 50 pregnant women agreed to listen to prenatal education music (experimental group); the other 50 pregnant women did not listen to prenatal education music (control group). Results in the anxiety scale show the experimental group that listened to prenatal education music had a lower score (36.64 points) than that of the control group which did not listen to prenatal education music (39.14 points). The results in the outpatient clinic satisfaction survey clearly show that letting the pregnant women listen to prenatal education music helped reduce their degree of dissatisfaction while waiting. Their positive feelings towards receiving medical treatment at the hospital were also enhanced. As for the measurement of objective physiological responses including vital signs such as the pregnant women's heart rate, systolic blood pressure, diastolic blood pressure and mean blood pressure, as well as placental function tests, both groups showed no significant differences. The research results show that prenatal education music can serve as a way for pregnant women to reduce anxiety while positively improving the degree of satisfaction for medical treatment received.

Keywords: Prenatal education music, Experience of receiving medical treatment, Anxiety experienced by pregnant women.

1 INTRODUCTION

According to the latest statistics from the Ministry of the Interior, the number of births and deaths in 2019 were 178,000 and 176,000 respectively, with a natural increase of 1,000 people. However, the number of births in July this year decreased by nearly 20% compared to the same period last year, while the cumulative number of deaths since the beginning of the year is 103,088, which is 10,714 more than the number of births. Simply put, the Golden Cross for population is expected to arrive in 2020, but the natural growth of the population has decreased by 10,000

people so far this year, according to the latest statistics (Department of Statistics, the Ministry of Interior, Total Number of Births). Today, Taiwan's birth rate is only 1.02 births per 1,000 people, the lowest in the world, which has posed a serious threat to the national security.

Many studies have proven that listening to refreshing and pleasant rhythmic music during pregnancy has a direct impact on the fetal limbic system and reticular formation, thus promoting development of the fetal brain as well as sensory integration. In addition, listening to melodious music helps expectant mothers to secrete specific substances that are good for health, promoting regulation of blood flow and stimulation of nerve cells. Furthermore, it improves blood supply to the placenta, and increases the essential properties of

---

*Corresponding Author

DOI 10.1201/9781003188513-17

the blood, which is very beneficial for fetal development and growth (Ye 2007). If we can find appropriate sound wave ranges from relevant studies, then customize exclusive headsets for expectant mothers with the most appropriate audio frequency music for the fetus, and make it specifically available in obstetrics and gynecology departments, expectant mothers will be more inclined to listen to the music longer for the benefit of the fetus. This can create additional value for the time spent waiting for a prenatal visit and increase willingness to wait.

Therefore, in this study, we invited expectant mothers who were waiting for prenatal visits to listen to prenatal music. Since the time would be spent waiting anyway, we creatively turned the negative emotions that would cause anxiety into a positive patient satisfaction that "they would be willing to wait, even if it took longer".

2 LITERATURE REVIEW

2.1 The effects of music on human bodies

(1) Professor Liu Zelun, the Chairman of the Prenatal Education Professional Committee within the Chinese Association for Improving Birth Outcome and Child Development, conducted an experiment with the Physiological Psychology Teaching Lab of Peking University. Fifty female undergraduate students were tested on changes in their microcirculatory blood flow while being exposed to loud music and soothing music. The results showed that loud music caused vasoconstriction in 96% of the subjects, while soothing music caused capillary dilation in 92% of the subjects (Ye 2007).

(2) In the fourth quarter of 2008, Wind Music started the collaboration with the Industrial Technology Research Institute (ITRI). The Center for Measurement Standards, ITRI was commissioned to conduct a study on soothing music, as carefully classified by Wind Music. The ITRI invited 25 subjects to listen to this music for 15 minutes, then measured individual brain wave changes. The ITRI found that the alpha waves of the subjects increased by 21% and the beta waves decreased by 13% after listening to the music classified by the Wind Music, as compared to those who were not listening to the music with their eyes closed. In addition, mixed music that had been randomly shuffled and chopped up was also played for the subjects; the results suggested that the alpha waves in the brain increased by 24% when they were listening to music with specific melodies and frequencies, as compared to the mixed music that was disorganized in terms of its genre, indicating that music created by Wind Music helped to balance emotions and relieve tension and stress (Industrial Technology and Information Monthly Journal 2015).

3 RESEARCH METHODOLOGY AND PROCESS

3.1 Research design

This study investigated whether providing prenatal music to expectant mothers during the waiting time for prenatal visits can be a way to reverse negative maternal anxiety. In this study, the use of alpha wave music as prenatal music was expected to provide a method for rapid reduction of anxiety during prenatal visits, where the Situational Anxiety Scale was used to assess the anxiety level of expectant mothers after listening to the provided prenatal music. Meanwhile, a structured questionnaire was used to collect data for evaluation of "patient satisfaction" which was adapted from the Taiwan Healthcare Indicator Series (THIS), including five major aspects: hospital environment and facilities, waiting time, staff attitude, medical procedures, and patient safety. Doing so helped to understand the patients' perceptions and degree of satisfaction.

3.2 Research methodology (research subjects, venue, process)

This study investigated whether the use of prenatal music could reduce the anxiety of expectant mothers who waited too long for their medical consultation appointments (Wang et al. 2006) in the study, a quasi-experimental research design was adopted. The research was conducted at an obstetrics and gynecology hospital in central Taiwan (Yeh et al. 2006), for a 10-month period starting from March 1, 2017 to December 31, 2017, with expectant mothers who had 10 complete prenatal visits done at the hospital during the study period as the research subjects. Those who underwent continuous prenatal visits at the hospital, from their first visit until they were 24 or more weeks pregnant, were given a prenatal music player set (over-ear headphones + MP3 music player) for each prenatal visit at the hospital, provided that they agreed to be tested on prenatal music education and signed a consent form after the professional health educator had explained to them how the prenatal music could benefit both mother and fetus. They were instructed to listen to the prenatal music for 30 minutes in the outpatient waiting area following by a break; those who listened to prenatal music more than three consecutive times were asked to complete the Situational Anxiety Scale and Patient Satisfaction questionnaires (Liu et al. 2008)

3.3 Research tools

The purpose of this study was to use prenatal music to reduce the anxiety level of expectant mothers who waited a long time for their prenatal appointment, while the effectiveness of the prenatal music education was evaluated by both the Situational Anxiety Scale and Patient Satisfaction questionnaires. The research devices included music players for prenatal music, brain wave measurements, the Situational Anxiety Scale questionnaire, and the Patient Satisfaction questionnaire.

4 RESEARCH FINDINGS AND ANALYSIS

4.1 Research data collection

In this study, the data was collected in a hospital's outpatient department. The research was conducted at an obstetrics and gynecology hospital in central Taiwan over a 10-month period starting from March 1, 2017 to December 31, 2017. A total of 100 valid samples were collected from expectant mothers who had undergone 10 complete prenatal visits at the hospital during the study period. Of these, 50 expectant mothers were in the experimental group who agreed to be tested with prenatal music education, and 50 were in the control group who didn't listen to the prenatal music. All of them were asked to complete both of the Situational Anxiety Scale and Patient Satisfaction questionnaires at 36 weeks of pregnancy

4.2 Basic information of the research subjects

4.2.1 Age
A total of 100 valid expectant mothers were enrolled in this study, of which 50 were in the experimental group who agreed to be tested with prenatal music education, and 50 were in the control group who didn't listen to prenatal music. Age distribution for the two groups shown in Table 1 (Age Ratio of Experimental Group to Control Group) was compared; the majority of the expectant mothers in the experimental group were 31-40 years old (56%), while the majority of expectant mothers in the control group were aged 21-30 years old (54%).

4.2.2 Educational background
In the experimental group, those who graduated with a university or college degree accounted for the largest proportion (68%), followed by senior high school or vocational school (28%), with junior high school being the least (2%). In the control group, the majority of them graduated from senior high school or vocational school (78%), followed by junior high school (16%), with university or college graduates being the least (2%).

4.2.3 Occupation
There was no significant difference in the distribution of occupations between the experimental and control groups, but the proportion of unemployed expectant mothers was higher in the experimental group (34%) than in the control group (26%).

4.2.4 Parity
There was no significant difference in the distribution of parity between the experimental and control groups, with first births accounting for the largest number in both groups (42% and 46%, respectively).

4.3 Subjective evaluation on the effectiveness of prenatal music in reducing expectant mothers' anxiety

The research subjects were required to complete the Situational Anxiety Scale questionnaire by describing their feeling when listening to the prenatal music, with special emphasis on their "current feelings" Each question of the scale has four options: "Not at all" "Somewhat" "Quite" and "Very" with scores of 1, 2, 3, and 4 respectively. In this study, 100 valid samples were completed, of which 50 expectant mothers were in the experimental group who agreed to be tested with prenatal music education and the other 50 expectant mothers were in the control group who didn't listen to prenatal music. The statistical results, as shown in Table 2, showed that the number of questions in which the expectant mothers in the experimental group felt positive emotions was significantly higher than that in the control group. As shown by the comparison of improvement rates for the experimental and control groups, the highest improvement rate was 15.97% for question No. 5, "I feel so relaxed now" followed by question No. 8, "I feel very satisfied now" at 12.59%. This indicates that listening to prenatal music significantly improves subjective perceptions of positive feelings such as relaxation and satisfaction in expectant mothers.

Ranked results on the situational anxiety scale for the experimental and control groups

4.4 Comparing the effectiveness of prenatal music on patient satisfaction

The statistical results of the "Patient Satisfaction Questionnaire" used in this study, as shown in Table 1, respectively shows the scores of the five major aspects (namely, hospital environment and facilities, waiting time, staff attitude, medical procedures, and patient safety), which helps to understand patients' perceptions and satisfaction. A total of 100 valid samples were completed, of which 50 expectant mothers were in the experimental group who agreed to be tested with prenatal music education and the other 50 expectant mothers were in the control group who didn't listen to prenatal music. The overall satisfaction score was 4.04 (out of 5), with 4.06 for the experimental group and 4.01 for the control group. Of all aspects, medical procedures (4.28) and patient safety (4.28) had the highest satisfaction scores in the experimental group. On the other hand, staff attitude (4.26) scored the highest in the control group, with waiting time (3.69) scoring the lowest in terms of patient satisfaction. The overall patient satisfaction mean value of the 50 expectant mothers in the experimental group who listened to prenatal music was 4.06, which was between "satisfied" and "very satisfied" For this group, the highest scores (4.28) were respectively for medical procedure and patient safety aspects, while the lowest score (3.57) was for the hospital environment and facilities. In the control group, the overall mean score of patient

Table 1. Ranked patient satisfaction results for experimental and control groups

Satisfaction Survey	Experimental Group (listened to prenatal music) Mean Value (N=50)	Ranking	Control Group (did not listen to prenatal music) Mean Value (N=50)	Ranking	Improvement Rate (%)
1. Hospital Environment and Facilities					
1. Air conditioning moderate (cold, warm), lighting good	3.44	18	4.12	9	−16.5
2. Cleanliness of floor and aisles	3.44	18	3.60	18	−4.44
3. Cleanliness of consulting room and medical instruments	4.18	13	4.10	10	1.95
4. Restroom cleanliness	3.2	20	3.28	20	−2.43
2. Waiting Time					
5. Waiting time for registration	3.86	15	3.84	16	0.52
6. Waiting time for medical consultation	4.30	5	3.42	19	25.73
7. Your consultation time with the doctor	3.84	16	3.88	15	−1.03
8. Waiting time for payment and medication collection	3.68	17	3.64	17	1.09
3. Staff Attitude					
9. Service attitude of staff dealing with payment collection and patient registration	4.20	12	4.26	2	−1.41
10. The doctor is kind and friendly	4.38	1	4.28	1	2.33
11. The nursing staff is kind and friendly	4.22	11	4.24	5	−0.47
12. The attitude of the pharmacist	4.30	5	4.26	2	0.93
4. Medical Procedures					
13. The doctor examined you carefully and explained your condition in detail.	4.36	2	4.24	5	2.83
14. The skill of the doctor	4.34	3	4.26	2	1.87
15. The nursing staff respect your privacy.	4.30	5	4.22	8	1.89
16. Why do you choose to have your prenatal visits at our hospital?	4.12	14	4.06	13	1.47
5. Patient Safety					
17. The nursing staff verified your name.	4.28	8	4.24	5	0.94
18. The health educator verified your name and reassured the examination items.	4.28	8	4.08	12	4.90
19. The pharmacist verified your name and informed you clearly when to take your medication.	4.24	10	4.10	10	3.41
20. How do you feel about having your prenatal visits here at our hospital?	4.32	4	4.04	14	6.93

Improvement Rate = (Mean value of those who listened to prenatal music − mean value of those who didn't listen to prenatal music) * 100/mean value of those who didn't listen to prenatal music

Table 2. Ranked results on the situational anxiety scale for the experimental and control groups

Question	Experimental Group (listened to prenatal music) Mean Value (N=50)	Ranking	Control Group (did not listen to prenatal music) Mean Value (N=50)	Ranking	Improvement Rate (%)
1.I feel calm now.	3.14	2	2.86	3	9.79
2.I feel safe now.	3.20	1	2.92	2	9.59
3.I feel tense now.	1.84	12	1.82	12	1.10
I feel anxious now.	2.16	11	2.02	11	6.93
I feel very relaxed now.	2.76	9	2.38	10	15.97
I feel angry now.	1.10	20	1.10	19	0.00
I'm worried that something unfortunate might happen.	1.20	19	1.10	19	9.09
I feel very satisfied now.	3.04	4	2.7	6	12.59
I feel scared now.	1.48	13	1.58	14	−6.33
I feel comfortable now.	2.90	7	2.66	7	9.02
I feel confident.	2.62	10	2.42	9	8.26
I think I'm highly neurotic.	1.38	17	1.34	18	2.99
I'm always on edge.	1.38	17	1.40	17	−1.43
I think I'm indecisive.	1.48	13	1.56	15	−5.13
I'm relaxed now.	2.78	8	2.60	8	6.92
I'm very satisfied now.	2.94	6	2.80	5	5.00
I'm worried now.	1.48	13	1.60	13	−7.50
I'm confused now.	1.42	16	1.50	16	−5.33
I feel steady now.	3.02	5	2.84	4	6.34
I'm having a great time now.	3.12	3	2.96	1	5.41

Improvement Rate = (Mean value of those who listened to prenatal music − mean value of those who didn't listen to prenatal music) * 100/mean value of those who didn't listen to prenatal music

satisfaction was 4.01, which was also in between "satisfied" and "very satisfied"with staff attitude aspect (4.26) scoring the highest and the waiting time aspect (3.69) scoring the lowest.

5   CONCLUSION AND SUGGESTIONS

The results of this study showed that listening to prenatal music had an impact on the subjective perceptions of expectant mothers. According to the results of evaluating improvement in expectant mothers' anxiety, the most significant improvement was for "I feel very relaxed now" at 15.97%, followed by "I feel very satisfied now" at 12.59%. This indicates that listening to prenatal music can significantly improve the subjective perceptions of positive feelings such as relaxation and satisfaction in expectant mothers. In addition, the results of the expectant mothers' patient satisfaction were evaluated during this study, showing that prenatal music can reduce their dissatisfaction while waiting for medical consultation and increase positive perceptions toward the medical experience.

REFERENCES

Department of Statistics, the Ministry of Interior, Total Number of Births.
Industrial Technology and Information Monthly Journal, 2015. Wind Music Uses Technology to Create a Cultural and Creative Stage, 280(2), pp. 57–63.
M.Y. Liu, F.C. Hsu, S.E. Kuo, S.T. Chung, 2008. Exploring the Patient Satisfaction, Chung Shan Medical Journal, 19, pp. 209–220.
C.C. Wang, Y.C. Lee, Y.M. Yu, 2006. The Study of Waiting Time of Out-patient Departments in a Regional Hospital, Cheng Ching Medical Journal, 2(1), pp. 59–65.
N.H. Ye, 2007. Prenatal Education Esoterica (4th ed.), Taipei County: Dashulin Publishing House.
C.S. Yeh, B.P.H. Lin, L.C. Yin, 2006. Factors Influencing Choice of Delivery Places Among Pregnant Women in Taichung Area, The Journal of Health Science, 8(2), pp. 144–156.

Smart Design, Science and Technology – Lam et al (eds)
© 2021 the Author(s), ISBN 978-1-032-01993-2

A comparative analysis on the evaluation index of circular economy in Taiwan and Japan

Ming-Fu Ho*
Ph.D Program of Mechanical and Energy Engineering, Kun Shan University, Tainan, Taiwan

Huann-Ming Chou
Department of Mechanical Engineering, Kun Shan University, Tainan, Taiwan

ABSTRACT: The development of a circular economy is an effective way to reduce resource consumption, reduce waste discharge, and protect the environment. On the other hand, Material Flow Analysis (MFA) is the basis for national planning of circular economy building.

Japan is an advanced country that introduced the Material Flow Analysis (MFA) indicators earlier. Japan passed the Basic Act on Establishing a Sound Material-Cycle Society in May 2000. With the said Act as the basis, the "Fundamental Plan for Establishing a Sound Material-Cycle Society" was set up. In addition, based on the MFA results, the government's future development goals were set. In the plan, three main indicators, namely, resource productivity, cyclical use rate, and final disposal volume, were adopted to disclose the situations for input, recycling, and output of resource in Japan's sound material-cycle society. Through the quantitative tracking of material flow in the region, various evaluation indicators of resource input, product output, and waste discharge in the region were accurately grasped, which facilitated the country's formulation of circular economy development policies and measures.

This paper shows some sketchy judgement on the advantages and disadvantages of Taiwan and Japan's circular economy developments in the recent decade through relevant data collection method and a comparative analysis of the empirical data of three material flow indicators, namely resource productivity; resource recycling; and final disposal volume. Following the comparative analysis for the main indicators of Taiwan and Japan's circular economy, it was found that Taiwan indeed falls far behind Japan in terms of circular economy development. Japan is an advanced country with the highest resource cyclical use rate and the fastest circular economy development. As the saying goes, "An advice from others may help one's defects", by learning Japan's circular economy development mode, it just may enable Taiwan to quickly enter the realm of a circular society.

Keywords: Circular economy, Material flow, Resource productivity, Cyclical use rate, Final disposal volume.

1 INTRODUCTION

Our country is small in area but densely populated, with a great shortage of natural resources. Thus, the development of a circular economy has become a key issue for the sustainable development of a country. The Japanese government has established the most complete legal system on circular economy in the world by combining it with its own practice while boldly adopting the Germany's. As a result, it has turned Japan into the country with the highest resource efficiency and the fastest growing circular economy (Zhongguancun International Environmental Industry Promotion Center 2005).

Since the three material flow indicators used in Japan's "Basic Plan for Establishing a Recycling-based Society" are in line with the 3R principle of circular

economy, it better reflects the macroeconomics of the circular economy. Hence, these three indicators are used in this paper as the index group to preliminary compare the evaluation index of circular economy in Taiwan and Japan.

2 KEY INDICATORS OF CIRCULAR ECONOMY IN TAIWAN AND JAPAN

In May 2000, Japan passed the "Basic Law for Establishing a Recycling-based Society". In accordance with this law, the "Basic Plan for Establishing a Recycling-based Society" was formulated, and Japan further set the government's development goals for the future based on the results of material flow analysis. The status of import, circulation, and export of resources in Japan's recycling-based society are

*Corresponding Author

DOI 10.1201/9781003188513-18

revealed in the plan through three main indicators: Resource Productivity, Cyclical Use Rate, and Final Disposal Amount.

Firstly, the calculation formulas and indicators are briefly described as follows:

Resource Productivity Indicator: Resource Productivity = GDP/Direct Material Input (DMI)

The GDP refers to the gross domestic production, whereas the DMI shows the material flow of imports, including the total amount of domestic natural resource extraction, imported natural resources and imported products. Thus, it indicates the GDP produced by the DMI of the domestic units.

Cyclical Use Rate Indicator: Cyclical Use Rate = Amount of Cyclical Use (Recycle + Reuse)/ Total Material Input (=Amount of Cyclical Use + Direct Material Input (DMI))

The amount of cyclical use refers to the amount of recyclable (recycled, regenerated and reused) resources encompassed by the DMI; the total material input is the amount of cyclical use and total DMI. From this indicator, we can understand the proportion of the total resources input in the country that are prepared for reusing and recycling, which is an indicator for the evaluation of material flow.

Final Disposal Amount Indicator: Final Disposal Amount = Final Disposal Volume of Waste

An output indicator of the material flow. The lower the final disposal amount, the more efficient an economy is on the resource utilization and the less pressure is put on the environment.

3 THE DATA AND RESULTS OF THE MAJOR INDICATORS FOR CIRCULAR ECONOMY IN TAIWAN AND JAPAN

3.1 Research design

3.1.1 Relevant data: Covering the year of 2006 to 2014 (Final Disposal Amount 2007–2014)

1. The GDP in Taiwan from 2006 to 2014 is shown in Table 1 as follows:

Table 1. Taiwan's GDP.

Year	2006	2007	2008	2009	2010	2011	2012	2013	2014
GDP (USD in millions)	376375	393134	416961	392065	446105	485653	495845	511614	530043

Source: IMF International Financial Statistics (IFS), National Income Statistics by the Directorate-general of Budget, Accounting and Statistics, Executive Yuan, R.O.C. (Taiwan), and National Bureau of Statistics of China (National Bureau of Statistics of China, the GDP of the main countries.), compiled by this study.

2. The DMI (Direct Material Input) in Taiwan from 2006 to 2014 is shown in Table 2 as follows:

Table 2. Taiwan's DMI (Direct Material Input).

Year	2006	2007	2008	2009	2010	2011	2012	2013	2014
DMI (in million ton)	432	394	386	355	409	392	386	385	394

Source: Report on Material Flow Indicators at National Level by the Environmental Protection Administration Executive Yuan, R.O.C.(Taiwan) (Report on Material Flow Indicators at National Level by the Environmental Protection Administration Executive Yuan, R.O.C.(Taiwan)), compiled by this study.

3. The amount of cyclical use in Taiwan from 2006 to 2014 is shown in Table 3 as follows:

Table 3. The amount of cyclical use in Taiwan.

Year	2006	2007	2008	2009	2010	2011	2012	2013	2014
The Amount of Cyclical Use (in million ton)	12.19	15.56	15.15	16.02	16.46	16.83	16.78	17.42	17.67

Source: Report on Material Flow Indicators at National Level by the Environmental Protection Administration Executive Yuan, R.O.C.(Taiwan) (Report on Material Flow Indicators at National Level by the Environmental Protection Administration Executive Yuan, R.O.C.(Taiwan)), compiled by this study.

4. The final disposal volume of waste in Taiwan from 2007 to 2014 is shown in Table 4 as follows:

Table 4. Final disposal volume of waste in Taiwan.

Year	2007	2008	2009	2010	2011	2012	2013	2014
Municipal Solid Waste (in million ton)	1.28	0.95	0.79	0.86	0.74	0.63	0.74	0.57
Business Waste (in million ton)	2.31	1.29	1.20	0.73	0.31	0.27	0.37	0.47
Total	3.59	2.24	1.99	1.59	1.05	0.90	1.11	1.04

Source: Report on Material Flow Indicators at National Level by the Environmental Protection Administration Executive Yuan, R.O.C.(Taiwan) (Report on Material Flow Indicators at National Level by the Environmental Protection Administration Executive Yuan, R.O.C.(Taiwan)), compiled by this study.

3.1.2 Results

The three indicators of resource productivity, cyclical use rate, and final disposal volume of waste in Taiwan are calculated and shown in Table 5 as follows:

Table 5. Key indicators of material flow analysis in Taiwan.

Year	2006	2007	2008	2009	2010	2011	2012	2013	2014
Resource Productivity (USD in millions/ in ten thousand ton)	871	998	1080	1104	1091	1239	1285	1329	1345
Cyclical Use Rate (%)	2.75	3.8	3.77	4.31	3.87	4.11	4.16	4.33	4.29
Final Disposal Amount (in million ton)		3.59	2.24	1.99	1.59	1.05	0.90	1.11	1.04

3.2 Japan's data and results

3.2.1 *Relevant data: Covering the year of 2006 to 2014 (Final Disposal Amount 2007–2014)*

1. The GDP in Japan from 2006 to 2014 is shown in Table 6 as follows:

Table 6. Japan's GDP.

Year	2006	2007	2008	2009	2010	2011	2012	2013	2014
GDP (USD in million)	4376005	4379749	4886963	5068059	5272943	5377426	5548566	5744195	5972119

Source: IMF International Financial Statistics (IFS), National Income Statistics by the Directorate-general of Budget, Accounting and Statistics, Executive Yuan, R.O.C. (Taiwan), and National Bureau of Statistics of China (National Bureau of Statistics of China, the GDP of the main countries.), compiled by this study.

2. The DMI (Direct Material Input) in Japan from 2006 to 2014 is shown in Table 7 as follows:

Table 7. Japan's DMI (Direct Material Input).

Year	2006	2007	2008	2009	2010	2011	2012	2013	2014
DMI (in million ton)	1591	1558	1492	1307	1364	1333	1362	1404	1389

Source: Annual Report on Environmental Statistics by the Ministry of the Environment, Government of Japan (Annual Report on Environmental Statistics by the Ministry of the Environment, Government of Japan 2017 Version.), compiled by this study.

3. The amount of cyclical use in Japan from 2006 to 2014 is shown in Table 8 as follows:

Table 8. The amount of cyclical use in Japan.

Year	2006	2007	2008	2009	2010	2011	2012	2013	2014
The Amount of Cyclical Use (million ton)	233	244	245	229	246	238	244	269	261

Source: Annual Report on Environmental Statistics by the Ministry of the Environment, Government of Japan, compiled by this study.

4. The final disposal volume of waste in Japan from 2007 to 2014 is shown in Table 9 as follows:

Table 9. Final disposal volume of waste in Japan.

Year	2007	2008	2009	2010	2011	2012	2013	2014
Municipal Solid Waste (in million ton)	6.35	5.53	5.07	4.84	4.82	4.65	4.54	4.3
Business Waste (in million ton)	20.65	16.47	13.73	14.36	12.58	13.25	11.76	10.5
Total	27	22	18.8	19.2	17.4	17.9	16.3	14.8

Source: Annual Report on Environmental Statistics by the Ministry of the Environment, Government of Japan (Annual Report on Environmental Statistics by the Ministry of the Environment, Government of Japan, 2017 Version.), compiled by this study.

3.2.2 *Results*

The three indicators of resource productivity, cyclical use rate, and final disposal volume of waste in Japan are calculated and shown in Table 10 as follows:

Table 10. Key indicators of material flow analysis in Japan.

Year	2006	2007	2008	2009	2010	2011	2012	2013	2014
Resource Productivity (USD in millions/in million ton)	2750	2811	3275	3878	3866	4034	4074	4091	4300
Cyclical Use Rate (%)	12.77	13.54	14.1	14.91	15.28	15.15	15.19	16.08	15.82
Final Disposal Amount (in million ton)		27	22	18.8	19.2	17.4	17.9	16.3	14.8

4 DATA ANALYSIS AND COMPARISON OF THE KEY CIRCULAR ECONOMY INDICATORS IN TAIWAN AND JAPAN

4.1 *Resource productivity*

Figure 1 below shows the difference in resource productivity between Japan and Taiwan. According to the data shown in Tables 5 and 10 above, Taiwan's resource productivity indicator was about 32% of Japan's in 2006, and it still made only 31% of Japan's in 2014. In general, Japan's resource productivity grew rapidly by 38% from 2007 to 2009, and increased slowly thereafter. In 2014, it increased by 56% compared to 2006. In Taiwan, resource productivity grew slowly from 2006 to 2014, which increased by 54% in 2014 compared to 2006. As the resource productivity indicator represents the effectiveness of natural resource utilization, it indicates that the natural resources productivity and utilization rate in Taiwan is at a low level compared to Japan.

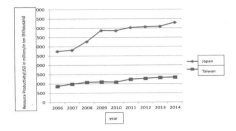

Figure 1. Comparison of resource productivity in Taiwan and Japan.

4.2 Cyclical use rate

Figure 2 below shows the difference in cyclical use rate between Japan and Taiwan. From the data shown in Tables 5 and 10 above, Taiwan's cyclical use rate was 2.75% in 2006, meanwhile Japan's was 12.77%. In 2014, it became 4.89% in Taiwan and 15.82% for Japan. The cyclical use rate represents the proportion of the total domestic resource input that are prepared for reusing and recycling; the higher the cyclical use rate, the higher the resource efficiency, and the lower the amount of waste disposal relatively.

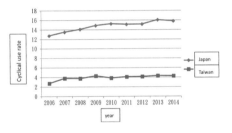

Figure 2. Comparison of the cyclical use rate in Taiwan and Japan.

4.3 Final disposal amount

Figure 3 below shows the changes of the final disposal amount between Japan and Taiwan. The lower the final disposal amount, the more efficient an economy is on the resource utilization, and the less pressure is put on the environment. From the data shown in Tables 5 and 10 above, the final disposal amount in Taiwan decreased by 71% in 2014 compared to 2006, while in Japan, it decreased by 74% in 2014 compared to 2006.

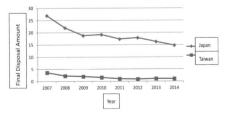

Figure 3. Comparison of the final disposal amount in Taiwan and Japan.

5 CONCLUSION AND DISCUSSION

The study found that Taiwan's resource productivity indicator was about 32% of Japan's in 2006, and it still made only 31% of Japan's in 2014. As the resource productivity indicator represents the effectiveness of natural resource utilization, it indicates that the natural resources productivity and utilization rate in Taiwan is at a low level compared to Japan. There are two reasons: (1) Most of the products require higher resource consumption and they are mostly OEM products, resulting in a lower product value. (2) Poor resource utilization techniques prevent the resources to be effectively utilized. In addition, the cyclical use rate in Taiwan was 2.75% in 2006, meanwhile it was 12.77% in Japan, and in 2014, it became 4.89% in Taiwan and 15.82% for Japan; it indicates that there remains considerable room for improvement in the reusing and recycling works in Taiwan. Since the cyclical use rate represents the proportion of the total resources input in the country that are prepared for reusing and recycling, therefore, the higher the cyclical use rate, the higher the resource efficiency, and the lower the amount of waste disposal relatively; this is also an important issue that Taiwan must face in developing the circular economy. In regards to the final disposal amount in Taiwan, it decreased by 71% in 2014 compared to 2006, while in Japan, it decreased by 74% in 2014 compared to 2006, indicating that Taiwan has made great progress in waste reduction, reusing and recycling, but we still need to continue putting in more efforts in the respective areas.

After comparing and analyzing the key indicators of the circular economy in Taiwan and Japan, the study showed that the respective area still requires much improvements. Whereas, Japan has become the country with the highest resource efficiency and the fastest growing circular economy. As the saying goes, develop the nation effectively by borrowing talent from abroad. By learning from Japan's circular economy development models, we may be able to transform our society into a circular economy within a short period of time.

REFERENCES

Annual Report on Environmental Statistics by the Ministry of the Environment, Government of Japan, 2017 Version.

National Bureau of Statistics of China, the GDP of the main countries.

Report on Material Flow Indicators at National Level by the Environmental Protection Administration Executive Yuan, R.O.C.(Taiwan).

Zhongguancun International Environmental Industry Promotion Center, 2005. Recycling Economy: Global Trends and Practices in China, Beijing: People's Publishing House.

Smart Design, Science and Technology – Lam et al (eds)
© 2021 the Author(s), ISBN 978-1-032-01993-2

The application of vertical greening for energy conservation and carbon reduction

Chen-Sheng Hsiung*
Ph.D Program of Mechanical and Energy Engineering, Kun Shan University, Tainan, Taiwan

Huann-Ming Chou
Department of Mechanical Engineering, Kun Shan University, Tainan, Taiwan

ABSTRACT: Greening is an important indicator of the modern urban living environment. It is also the most natural way to solve the carbon dioxide problem, which is conducive to alleviating the crisis of global warming. The air quality in Taiwan is aggravating due to the energy policy at present, and there seems to be no better solution in the short run. As a result, this highlights the importance of greening. This study is intended to extend the concept of vertical greening to communities and families by making use of the greening wall technique and characteristics of different plants such as vines, flowers, and grass, in order to enhance the results of the air purification. Hopefully, this will improve the gradual worsening air pollution problem. Besides, adding the form of ecological green life, the goals of energy conservation and carbon reduction in small areas can be achieved, thereby our daily life will become greener, more organic and environmentally friendly.

Keywords: Greening, Vertical greening, Greening wall, Energy conservation and carbon reduction.

1 INTRODUCTION

1.1 Research background and motivation

As a result of the continuous rising of the earth's temperature, the natural structure that enables the normal operation of the earth is destroyed, affecting the living environment of people and the survival of other species. The development of technologies and the waste of resources by human have brought damages to the environment. How to repair the damaged natural structure is the greatest responsibility of mankind.

A news report on July 21, 2020 stated that, according to a research published in the Nature Climate Change journal, climate change is causing negative impact on polar bears, making them difficult to find food, which leads to their starvation and death. It is expected that polar bears may become extinct within 80 years. Even if the rising temperature of the earth is maintained within 2.4 degrees, it may only delay the extinction of polar bears for a short period of time (TVBS News (CNA News) 2020).

Siberia, located in the Arctic Circle, experienced record-high temperature in June 2020. Under the attack of the heat wave, the worst wildfire in Siberia was triggered. As of July 27, the wildfire in Russia has spread out to cover an area of 20 million hectares, which is equivalent to 5.5 times the area of Taiwan, of which 11 million hectares are forests (Greenpeace website 2020).

The two examples mentioned above highlight the serious problems caused by the environmental changes around the world. Therefore, it is urgent for mankind to find solutions to these environmental problems. This paper will discuss the issues of energy conservation and carbon reduction in life from the perspective of global environmental changes and apply the idea of energy conservation and carbon reduction to vertical greening of residential landscaping. The research method adopted by this paper is based on the discussion of literature data and content analysis.

1.2 Research objectives

Due to the COVID-19 pandemic in 2020, many countries and regions have closed their cities to block the spread of the virus, ensuring an effective control of the epidemic and the number of infected people. By closing down cities and countries around the world, the global environment accidentally received a relief, and damages caused by humans were reduced. As a result, damages to the ocean, beaches and the sky due to the gathering of people were minimized, leading to less water pollution, air pollution and light pollution. Many animals start to enjoy the beautiful natural environment. Owing to the impact of the COVID-19

*Corresponding Author

DOI 10.1201/9781003188513-19

epidemic, we once again realize the importance of the ecosystem, waste reduction, greening and health.

Vertical greening has been adopted by many countries in European and America for a long time, and there are several cases that we can learn from. This study aims to introduce the concept and the techniques of vertical greening into home environment, allowing vertical greening to be applied effectively and rapidly in home environments based on the aspects of everyday life. It is hoped that such change in the microenvironment can improve the efficiency of energy conservation and carbon reduction.

This study hopes to reach the following objectives:

1. Allowing the concept of vertical greening to go deeper into home design.
2. Transforming Taiwan's collective housing model and implement the concept of greening into architectural design.
3. The concept of environmental protection is adopted to make greening closer to everyday life, making life more natural, environmentally friendlier and healthier.

2   LITERATURE REVIEW

2.1   *The impact of global environmental change on ecosystem*

Due to the rapid increase in population and the advancement of technology, significant amount of carbon dioxide (CO_2), nitrous oxide (NO_2), and methane (CH_4) have been produced. The huge emission of these greenhouse gases has destroyed the surface structure of the earth, causing the temperature of the earth's surface to rise, which leads to climate change.

2.1.1   *The impact of greenhouse effect*
According to data from the National Aeronautics and Space Administration (NASA) and the National Oceanic and Atmospheric Administration (NOAA), 2018 is the 4th hottest year under the global warming trend, which also ranks the 4th in the year with the highest global surface temperature since 1880. Since the 80s, the global average surface temperature has increased by about 1°C. In the past few years, Taiwan continues to set new record for the highest temperature; the average temperature in 2019 reached a new high.

According to the Intergovernmental Panel for Climate Change (IPCC) of the United Nations, the global temperature in 2100 will be 0.9°C to 3.5°C higher than that in 1990. This implies that more natural disasters will occur due to abnormal climate, more species will face extinction, the food chain of various species will be destroyed, and humans will face more disasters.

Scientists predict that each time carbon dioxide is released into the atmosphere in multiples, the average temperature on the earth will be increased by 1.5°C to 4.5°C. If the emission of greenhouse gases such as carbon dioxide, methane, etc. is not controlled, the earth's temperature may be increased by 4°C or higher, creating challenges for mankind. On the other hand, if the emission of greenhouse gases is effective controlled, the increase in the earth's temperature can be controlled at around 1°C.

2.1.2   *Control the earth's temperature to stop natural disaster from abnormal climate*
a. Flood
As the global average temperature rises, the sea temperature also rises, leading to the melting of ice in the Arctic regions and rising of the sea level. It is projected that in 2100, the sea level will rise by 38 cm to 56 cm as compared to that in 1990. As a result, many cities on the surface of the earth will be submerged, or their areas will be reduced due to land loss. In other words, the occurrence of floods will be more frequent around the world. In the past two years, flooding disasters have occurred in many countries, such India, China, Japan, South Korea, etc.
b. Forest Fire
In June 2020, a forest fire broke out in Siberia, Russia, burning more than 4.6 million hectares of forest. In 2019, forest fire occurred in the Amazon rainforest of Brazil. From September 2019 to February 2020, a forest fire that continued for six months in Australia killed 1 billion animals. The fire burned an area of 17.19 million hectares, which is 4.7 times the area of Taiwan. In August 2020, a forest fire occurred in California, USA. The fire burned an area of 10,000 acres, causing significant damages to many local residents. These forest fires are all caused by climate change, making the areas that are already prone to wildfires due to dry weather even more vulnerable to forest fire.

2.1.3   *End of the World*
According to the analysis the earth will no longer be suitable for humans to live on if its temperature is increased by 6°C, which may lead to the end of the world (Environmental Protection Administration, Executive Yuan official website).

2.2   *Develop green buildings in response to environmental changes*

2.2.1   *Global green building evaluation system*
Green building has been designed in Europe, America and Asia for a while. Therefore, systems have been developed for evaluating green buildings, reaching a peak after 2000. As of 2018, the green building evaluation system has been adopted by 38 countries around the world.

With the help of the green building evaluation system, buildings can be designed based on the concept of environmental protection. By adopting more natural light and ideal traffic into the design, waste of energy can be minimized to protect the earth's resources, reducing the heat island effect to achieve energy conservation and carbon reduction.

2.2.2 Taiwan green building evaluation system – EEWH

Taiwan has developed a green building evaluation system that is suitable for assessing buildings in the subtropical countries. The system contains 9 major indicators to set a quantitative benchmark for evaluation in 4 key areas including ecology: biodiversity, greening volume, base water retention; energy saving: daily energy saving; waste reduction: CO_2 reduction, waste reduction; and health: improvement of indoor environment, water resources, sewage and waste. Based on these 9 major indicators, buildings are evaluated to see whether they are in line with the current requirements for environmental protection. Due to the rising trend, there is one important concept that must be understood; smart buildings are not equivalent to green buildings. It is not our intention to design a building that has the convenience of a smart building but consumes more electricity or energy due to the adopted technologies, deviating from the purpose of a green building which is environmental protection and carbon reduction (Lin 2016). This will be another issue for discussion.

2.3 Benefits and development trend of green wall

2.3.1 Development trend

Vertical greening of building structures has been effectively developed in recent years. Several successful and innovative ideas have been seen across Europe and America or even in Asian countries such as Singapore and Japan. Through the promotion of related policies, such as roof greening, by the governments in various countries, the greening of building has become a trend. For example, the Tokyo City Government of Japan first formulated laws and regulations on roof greening and wall greening, forcing the central and the local governments to promote greening activities and forming a trend. In Taiwan, vertical greening through various studies has achieved several achievements.

Green walls are often used in the construction industry. Typically, they are applied by incorporating wall greening or roof greening into the design. In Taiwan's green building evaluation system, green wall is also included as one of the items for assessment. Since part of the greening area of the building will be removed when designing the traffic routes, the use of roof greening and vertical greening of walls can recover the greening area. Green wall is a greening method that has been used quite often in architectural design. It can be divided into two categories, pixel-type and vine-type.

The pixel-type green wall is formed by placing the planted plants into the designated holes or containers to carry out the design. It is also called the Early Green Façade Type green wall, which is commonly used in vertical greening in many offices, public spaces, or homes. The walls can be designed and arranged in advance, and then assembled on site.

The vine-type green wall is formed by planting vine-like plants on designated walls or placing them on assembled frames. By utilizing the climbing nature of vine-like plants, the plants can grow along the frames to achieve the effect of painting the wall. It is also called the Future Green Façade Type green wall. This type of green wall is relatively easy to maintain. As long as it is properly watered, it will grow naturally on its own. On the other hand, the pixel-type green wall requires special attention in term of the applied media and watering because it is formed by transplantation. In addition, factors such as environmental protection and the quality of water sources must be taken into consideration, in order to avoid waste during maintenance, fulfilling the purpose of environmental protection, energy conservation and carbon reduction.

2.3.2 Benefits

The installation of green walls can bring functional and sensory benefits.

a. For functional benefits, green walls can achieve the goal of energy conservation and carbon reduction, reducing the Urban Heat Island Effect. Green walls can help buildings block the radiant heat generated by the sun. The evaporation of water from leaves of the plants and in the soil can slightly decrease the temperature at the site, reducing the heat island effect. According to the studies conducted by the United Nations Environment Programme, if buildings in cities achieve 70% roof greening and vertical greening, the photosynthesis process carried out by plants can reduce carbon dioxide concentration in the cities by 80%. In this case, the problem of heat island effect can be solved. Therefore, reducing the concentration of carbon dioxide will be the major benefit of installing green walls. With regards to environmental protection, the benefits of installing green walls include reducing air pollution, reducing noise, and forming stepping stones—the creation of "microhabitats" in the environment to favor biodiversity. As for the economic benefits, green walls can reduce the cost of electricity for air conditioning systems and increase the real estate value (Green Wall Handbook 2016).

b. Sensory benefits improve life aesthetics, reduce everyday stress, visual satisfaction to purify body and mind, and enlighten the soul.

3 ANALYSIS ON THE DECREASING OF INDOOR TEMPERATURE BY VERTICAL GREENING

The greening coverage rate in Singapore is about 50%, while the greening coverage rate in Taipei is 5%. This suggests that the number of trees in Singapore is 9 times more than that in Taipei and almost half of the sights of the people in Singapore include green plants. Trees can provide shades for people to keep them away from sunshine and rain. They can also adjust the tropical climate, reduce industrial pollution and improve the highly developed urban environment. A green city can replenish oxygen, reduce carbon, absorb solar radiation, and maintain water balance, thereby improving the habitability of the city through factors such as natural environment, medical care and health.

Studies have shown that insulation on the outer surface of a building can greatly reduce the electricity costs for air conditioning systems. According to the study conducted by the Hsi Liu Environmental Greening Foundation, taking a top-floor apartment with an area of about 132 m^2 as an example, the apartment with roof greening can save 93 USD in electricity costs for the air conditioning system in summer as compared with apartment without roof greening, which is equivalent to reduce 515 kg of carbon dioxide emission. For building walls that are exposed to sunlight, if the build has green walls, similar energy-saving effect can be achieved (Green Wall Handbook 2016).

4 CONCLUSION

Climate change has become a reality. The best we can do is to find out how to mitigate and overcome the disasters it brings. People need to constantly respond and improve their lifestyles in order to reduce occurrence of disasters. Many countries have also worked hard to formulate agreements and conventions for people to follow through various methods in order to achieve energy conservation and carbon reduction, protecting the natural environment and resources.

In daily life, with this concept, greening is introduced into communities and homes. By adopting vertical greening design and using natural plants to improve the micro-environment and temperature, the final goal of energy conservation and carbon reduction can be gradually achieved.

With Taiwan's economic growth and the increase in power consumption for households and industries, and the government's energy policy towards non-nuclear homes, the Taichung Thermal Power Plant and the Datan Thermal Power Plant are joining the power generation systems of Taiwan. According to the power generation ratio provided by Taipower, the proportion of thermal power generation increased year by year. The thermal power generation of the Taipower systems accounted for 79.18% in 2019. Since nearly 80% of the power generated in Taiwan is from thermal power plants, the worsening of air pollution in Taiwan is inevitable.

From the above data, it is obvious that the thermal power generation is the major source of electricity in Taiwan, making the air pollution problem in Taiwan to become more and more severe. Therefore, topics on reducing electricity consumption for domestic use, reducing the greenhouse effect, lowering indoor temperature to minimize the electricity consumption for air conditioning systems and achieve the goal of energy conservation and carbon reduction, and implementing designs from household greening to green buildings will become quite important.

REFERENCES

Environmental Information Center, January 03 2020.
Environmental Protection Administration, Executive Yuan official website.
Green Building Evaluation Manual-Eco-Community, September 2019. Architecture and Building Research Institute, Ministry of the Interior.
Green Wall Handbook, February 2016. Architecture and Building Research Institute, Ministry of the Interior. Taiwan Power Company.
Greenpeace website, 2020/7/29.
IPCC Report: Climate Change 2013: The Physical Science Basis, AR5-WG1 FigSPM-07a.
H.T. Lin, 2016. Technology Guides for Green Building Design, Chan's Arch-publishing Co., Ltd.
NASA, earth observatory, 2018 Was the Fourth Warmest Year, Continuing Long Warming Trend.
TVBS News (CNA News), 2020/7/21.

Smart Design, Science and Technology – Lam et al (eds)
© 2021 the Author(s), ISBN 978-1-032-01993-2

An investigation on the variation of carbon footprint and fertilizer content after treating food waste with black soldier fly composting

Li-Chin Chang*
Ph.D Program of Mechanical and Energy Engineering, Kun Shan University, Tainan, Taiwan

Huann-Ming Chou
Department of Mechanical Engineering, Kun Shan University, Tainan, Taiwan

ABSTRACT: According to the statistics of the Food and Agriculture Organization of the United Nations (UNFAO), about 1/3 of the world's food is wasted and discarded as organic waste every year, resulting in 4.4 $GtCO_2$ eq of carbon emission. If the waste cannot be properly reutilized, the Earth will be polluted, leading to catastrophic damages. In order to reduce pollution to the environment and carbon emission, this study proposes the Black Soldier Fly open-air composting method for treating organic wastes, such as food wastes. To evaluate the efficiency and convenience of the Black Soldier Fly composting, the carbon emission and fertilizer content of this method were estimated based on the experimental data, and the results were compared with those of other methods. The proposed method not only can improve the efficiency of traditional composting, but also can eliminate the difficulty in breading Black Soldier Fly since the purpose is not collecting insect protein. According to the experimental results, the method proposed in this study can be easily operated in farms as well as in communities. Moreover, the method is suitable for converting the problematic food wastes in small batches into organic fertilizer with high fertilizer content.

Keywords: Black soldier fly, Compost, Fertilizer content analysis, Carbon emission, Greenhouse gases.

1 INTRODUCTION

According to statistics from Taiwan's Environmental Protection Administration, general waste contained more than 1/3 of food waste (publication edited by the Environmental Protection Administration, Executive Yuan 2016). The current annual generation of food waste was about 500,000 tons (Environmental Protection Administration, EY-Statistics 2020). In the past, food waste could be used for feeding pigs, so it was not a big problem. From 2019 onwards, in order to prevent the spread of African swine fever, feeding food waste to pigs has been banned, and thus a lot of food waste was mixed with general waste and burned in incinerators, causing serious problems of air pollution and the release of contaminants such as dioxins, heavy metals, etc. Compared with open-air composting, the biological composting method can reduce the global warming potential by 47 times (Mertenat et al. 2019). Therefore, the use of black soldier flies to convert food waste can reduce more carbon emissions.

2 BLACK SOLDIER FLY COMPOSTING AND CARBON EMISSION ESTIMATION

In this study, the food waste was placed in a compost area where insect protein was not collected and black soldier flies were used to accelerate the food waste composting process at a speed of about 2-4 times faster than the general traditional composting method. This method was implemented by digging down to a depth of 50 cm to form a one-cubic-meter hole in the field, and then putting 20 liters of food waste into it regularly every week. After dumping the food waste, the hole was lightly covered with thin layers of soil or weeds to prevent mosquitoes and flies from breeding and to prevent it from producing odor, by which organic fertilizer was to be harvested in one month. Except for the initial breeding insects placed in the compost area, female black soldier flies were then attracted by food and the smell of black soldier fly larvae to lay their eggs in the compost, thereby forming a biological compost ecosystem. No additional black soldier flies or eggs were required. By observing the actual ecosystem, besides black soldier flies, there were also other bugs such as red earthworms, black earthworms, millipedes, centipedes, etc., living together and cooperating to decompose the food waste,

---

*Corresponding Author

DOI 10.1201/9781003188513-20

Table 1. Greenhouse gas emissions of various food waste treatment methods.

Item	kg/ton	References
Black soldier fly	30.2kg	(Salomone et al. 2017)
Incineration	99.3~510	(Chen et al. 2013a, 2013b; Liao 2014)
Pig feeding	55	(Chen et al. 2013a, 2013b; Liao 2014)
Aerobic composting	33.4~110.1	(Chen et al. 2013a, 2013b; Liao 2014)
Anaerobic composting	0~1073	(Chen et al. 2013a, 2013b; Liao 2014)

Note: 1. Aerobic composting data differed due to "good operation" vs "bad operation".
2. The data of anaerobic composting differed significantly; the key factor of which was whether the gas produced after treatment had been recovered or not.

Table 2. Analysis results of various solid fertilizers.

Item		Food waste manure	Abattoir sludge manure	Food waste compost	A-hou City Organic Fertilizer	Limit Value 1	Limit Value 2
Organic matter	%	90.1	82.64		73	–	–
Nitrogen	%	3.596	3.271	1.52	2.6	–	–
Phosphorus	%	1.561	2.266	0.02	1.9	–	–
Potassium	%	2.384	2.801	0.01	1.4	–	–
Calcium	%	0.157	1.039	0.026			
Magnesium	%	0.696	0.938	0.01			
Iron	mg/kg	4056	4693				
Manganese	mg/kg	175.3	208.9				
Copper	mg/kg	36.22	**90.21**	1.7	47	<=100	<=100
Zinc	mg/kg	162.7	**434.8**	2.2	222	<=250	<=500
Cadmium	mg/kg	0.075	0.286	0.04		<=2	<=2
Chromium	mg/kg	82.21	34.46	0.73	16	<=150	<=150
Nickel	mg/kg	**41.65**	17.47	0.29	5.9	<=25	<=25
Lead	mg/kg	5.661	13.06	0.13	6	<=150	<=150
Arsenic	mg/kg	1.796	4.006		1.2	<=25	<=25

Note: According to the regulations on fertilizer product specifications, related values of livestock manure compost and miscellaneous compost shall not exceed Limit Value 2, and values of the remaining solid compost shall not exceed Limit Value 1.

by which organic fertilizer with rich fertilizer content was produced. Moreover, continuous feeding also provided black soldier flies with sufficient food, through which a black soldier fly ecosystem could be maintained. This method is easy to operate and suitable for regular decomposition of a small amount of food waste in the compost area. Apart from being faster and more energy efficient than general organic composting, its advantages also include no odor, no breeding of mosquitoes and flies, and no methane production.

By treating 20 liters of food waste per week, which is equivalent to about one ton of food waste per year, this study estimates and compares the greenhouse gas emissions of various food waste treatment methods such as black soldier fly, incineration, pig feeding, aerobic composting, anaerobic composting, etc. According to the research results, the global warming potential was 30.2 kg CO_2 eq (Salomone et al. 2017) when the black soldier fly method was used to treat 1 ton of food waste. In addition, by using the emission factors provided by IPCC to estimate the GHGs emissions of other food waste treatment methods including incineration, pig feeding, aerobic composting and anaerobic composting (Chen et al. 2013a, 2013b; Liao 2014), the greenhouse gas emissions of one metric ton of food waste treated with the above-mentioned methods were estimated and compared as shown in Table 1. It was found that the black soldier fly food waste treatment method had lower greenhouse gas emissions, which means it had a great advantage in term of carbon footprint.

3 FERTILIZER CONTENT COMPARISON BETWEEN BLACK SOLDIER FLY ORGANIC MANURE FERTILIZER AND TRADITIONAL COMPOST

Not many people use cooked food waste for composting because of its high water content, odor and mold breeding properties. Pierre Loisel was a pioneer in using recycled food waste for composting. He believed that with the diversification of sources and ingredients of recycled food waste, the fertilizer content of the compost produced would also be richer. Compared with the organic manure fertilizer harvested from food waste after being treated with black soldier flies, the cooked food waste compost possesses diversified fertilizer content and can be used for plants directly. This study compares the solid fertilizer content of various organic fertilizers, and the results are shown in Table 2.

Explanation of Results:

Sample 1: The organic manure fertilizer produced from food waste treated with black soldier flies. The food waste source was the cooked food waste collected from food courts of shopping centers (Lin 2016). Please refer to "Limit Value 1" for the fertilizer content standards.
Sample 2: The organic manure fertilizer produced from abattoir sludge treated with black soldier flies (Lin 2016). Please refer to "Limit Value 2" for the fertilizer content standards.
Sample 3: The food waste compost from Mr. Pierre Loisel's seaside farm (Soil test report of Mr. Pierre Loisel's seaside farm). Please refer to "Limit Value 1" for the fertilizer content standards.
Sample 4: The commercial compost produced by Pingtung County's Nanzhou District Farmers' Association—the sample was livestock manure compost with its source coming from cow manure, pig manure, chicken manure, wood chips, etc., collected in Pingtung area (Pingtung County's Nanzhou District Farmers' Association). Please refer to "Limit Value 2" for the fertilizer content standards.

4 DESCRIPTION OF RESULTS

Sample 1: The food waste organic manure fertilizer contained high organic matter, nitrogen, phosphorus, and potassium, and contained nickel which exceeded the regulated standard. It implied that the stainless steel tableware and kitchen utensils used in these restaurants would release nickel, and highlighted the food safety issue of Taiwan's eating-out culture. Most of the stainless steel tableware may be substandard and likely to release nickel. This is why the heavy metal— nickel discharged by the black soldier flies after eating these ingredients exceeded the standard value. The organic manure compost produced by the black soldier fly larvae contained different fertilizer content because of different food sources. The heavy metal content in the feeding materials would directly affect the heavy metals contained in the black soldier fly larvae (Diener et al. 2015, Proc et al. 2020), and thus the heavy metals in their manure would also exceed the standard values. The problem did not lie in the black soldier flies, but in the food sources.

Sample 2: This sample contained the highest organic matter, nitrogen, phosphorus and potassium. The values all met the Limit Value 2 standards for livestock manure. But the measured values of copper and zinc were relatively high as copper sulfide and zinc oxide had been added to the feed to increase the efficiency of animal growth.

Sample 3: Mr. Pierre Loisel had been collecting food waste for organic fertilizer in Sanzhi District for years. Although the fertilizer content seemed lower, extra liquid fertilizer had been collected, and it was completely decomposed and harmless to the plants.

Sample 4: The fertilizer content of the commercial poultry manure fertilizer produced in Nanzhou District (with a brand name of Ahou) was moderate and stable.

It was found that the organic manure fertilizer harvested from food waste after being treated with black soldier flies contained high fertilizer content, yet relatively speaking, the safety of food sources should be noted.

5 CONCLUSION

Compared with composting, the application of black soldier flies to the treatment of food waste manifested higher efficiency and produced organic fertilizers with higher fertilizer content. In addition, it produced a lower carbon footprint than composting, pig feeding, etc., in terms of greenhouse gas emissions. But the problem was that the leftover human food eaten by black soldier flies would affect its test values and also reflected the issue of food safety in human world. It was found that heavy metal pollution in relation to tableware and livestock had become a potential risk in human life.

REFERENCES

H.L. Chen, Y.Z. Liao, Z.F. Xu, M.R. Zhou, 2013a. Greenhouse gas emissions from food waste recycling and reuse calculated by using different emission coefficients, Dissertation of the Department of Soil and Water Conservation, National Chung Hsing University.

H.L. Chen, Y.Z. Liao, Z.F. Xu, M.R. Zhou, 2013b. Estimating the GHGs Emissions of Different Food Waste Treatment Methods through IPCC Measures, Dissertation of the Department of Soil and Water Conservation, National Chung Hsing University Journal of Soil and Water Conservation, 45 (1): 457–464.

S. Diener, C. Zurbrügg, K.Tockner, 2015. Bioaccumulation of heavy metals in the black soldier fly, Hermetia illucens and effects on its life cycle. Journal of Insects as Food and Feed, 1(4), 261–270.

Environmental Protection Administration, EY-Statistics 2020. National General Waste Generation on the website of Environmental Protection Administration, Executive Yuan.

Y.Z. Liao, 2014. Research on the Evaluation of the Optimal Treatment of Food Waste through Greenhouse Gas Emission Estimation, Dissertation of the Department of Soil and Water Conservation, National Chung Hsing University, P1–106.

X.R. Lin, October 22, 2016. Black Soldier Fly Breeding Workshop, Test report provided by Pastor.

A. Mertenat, S. Diener, C. Zurbrugg, 2019. Black Soldier Fly biowaste treatment – Assessment of global warming potential, Waste Management Volume 84, 1 February 2019, pp. 173–181.

Pingtung County's Nanzhou District Farmers' Association, Soil test report of A-hou City Fertilizer provided by Nanzhou Farmers' Association.

K. Proc, P. Bulak, D. Wiłcek, A. Bieganowski, 2020. Hermetia illucens exhibits bioaccumulative potential for 15 different elements – implications for feed and food production. Sci Total Environ 723:138125

Publication by the Environmental Protection Administration, Executive Yuan, 2016. 2013-2015 Commissioned Project for the Analysis of Sample Composition of General Waste before Final Disposal (2014, 2nd year) pp. 73.

R. Salomone, G. Saija, G. Mondello, A. Giannetto, S. Fasulo, D. Savastano, 2017. Environmental impact of food waste bioconversion by insects: application of life cycle assessment to process using Hermetia illucens, Journal of Cleaner Prodution, 140(2): 890–905.

Soil test report of Mr. Pierre Loisel's seaside farm.

Smart Design, Science and Technology – Lam et al (eds)
© 2021 the Author(s), ISBN 978-1-032-01993-2

Solar heat storage barrel steady structure aseismatic and topology optimization method

Jui-Chang Lin*
Professor, Department of Mechanical Design Engineering, National Formosa University, Taiwan

Cheng-Jen Lin
Assistant Professor, Vehicle Engineer of National Formosa University, Taiwan

Po-Yen Lin & Iftikhar Ali
MS Scholar, Department of Mechanical Design Engineering National Formosa University, Taiwan

ABSTRACT: This research paper analyzed the seismic resistance performance of the solar heat storage barrel structure through seismic acceleration. The water storage barrel is placed on the M-type tripod, which must bear 500 kg of static water weight. Seismic analysis has two parts. Static analysis of the earthquake acceleration is 130gal (5 weak earthquakes), 225gal (5 strong earthquakes), 390gal (6 weak earthquakes) and 444 gals (6 strong earthquakes). The dynamic diachronic analysis was designed according to the seismic building codes to capture the triaxle acceleration of the Mino earthquake. Finally, the original structure is used to optimize the volume to achieve the new structural design.
   This study mainly used Abaqus finite element analysis software for simulation. The bracketed material is stainless steel angle steel (SUS304) with a thickness of 1-3mm. The primary analysis is divided into the original structure and the topology structure. Without reducing the rigidity and strength of the system, reducing the cost of the design is achieved. The simulated seismic performance results revealed that the original structure could withstand 444 gals of seismic acceleration. The overall structure stress does not go beyond the material yield point designed for safety structure. Finally, this study also optimized the topology analysis to reduce the volume by 10%, and even with the same load and earthquake resistance. The research results show that the optimized topology structure has a weight reduction of 10%, a maximum stress drop of 16.8MPa and a maximum strain reduction of 0.513 compared with the original design. The dynamic diachronic analysis results show that the original structure's maximum stress is 285.4MPa, and the maximum pressure of the topological structure is 289.5MPa. Both designs can withstand the strength of the Mino earthquake (444 gals). We provide a reference for the improvement of the manufacturer's design and experimental verification through this research and analysis.

Keywords: Abaqus, Finite Element Analysis, Earthquake Resistance, Topology Optimization, Building Seismic Code.

1   INTRODUCTION

This research uses Abaqus to simulate static structural analysis and dynamic seismic duration analysis. The static analysis of the structure was carried out by calculating the horizontal seismic force and the vertical seismic force using the Ministry of Interior Building Seismic Design Code. The dynamic analysis captures the chronological acceleration analysis of the Mino earthquake. Finally, the topological optimization results provide the manufacturer's new structure design and meet the cost reduction requirements while also providing product analysis data and experimental data to ensure product safety.

In 2014, Cai Kunlin[2] conducted an analysis through Tosca, the internal modeling group of Abaqus, and took the gantry machining center machine as research and discussion, and optimized the design for each component so that the overall structure weight, stress, and displacement reached the best topological results. The final results showed the value reduced by 9.56%, the pressure reduced by 36.07% and the deformation reduced by 25.27%. The structural design has significantly improved.

In 2014, Hong Lifen[1] used Abaqus to simulate the topology optimization design of CNC internal and external cylindrical grinders and selected the outer diameter grinding wheel headstock, grinder frame, and base for topology design analysis. The results showed that the overall material optimization was achieved

---

*Corresponding Author

by reducing the overall 5%. As a result, the stress is reduced by 0.871%, and the strain is reduced by 1.2%.

In 2008, Zhang Rongci[2] used Abaqus software to include the discussed structural parameters in the design optimization variables and optimized the topology, size, and shape of the bridge structure. The research first defined the overall with topology optimization. The position of the part is then optimized for shape and size. The variables considered are the cross-sectional area and the cross-sectional dimension of the region. Simultaneously, the corresponding allowable stress and maximum bearing capacity of the structure need to be considered to design and obtain the optimized bridge structure.

In 2019, Yang Jiawei [3–7] used Abaqus analysis to simulate horizontal water towers' structural impact caused by seismic forces and typhoon winds and studied seismic forces. We designed horizontal and vertical seismic forces with static simulations for structural analysis. The results showed that the structure did not exceed the material yield strength under a magnitude seven earthquake. The typhoon designed horizontal and vertical winds according to wind resistance design specifications. The results showed that the structure could withstand strong winds of magnitude 17 and finally reduced the material's volume by 10% through optimization analysis. It can still resist a magnitude seven earthquake and a magnitude 17 strong wind.

2 EXPERIMENTAL EQUIPMENT

This study used Abaqus finite element software. The analysis results obtained the maximum strain value and stress, then measured and verified through the strain gauges. DCS-100A software is used to capture and measure experiment data. The rest of the measuring equipment and software showed in Table 1–2 and Figure 1.

Table 1. Summary of used software.

Items	Software
1	Abaqus CAE
2	Abaqus Topology
3	DCS-100A

Table 2. Summary of measuring equipment.

Items	Equipment name	$$
1	Uniaxial strain gauge	KFGS-5-120-C1-11L1M2R
2	Connection unit	U-124
3	Unique adhesive for strain gauge	CC-33A
4	Computer control unit	EDX-10A
5	Strain measurement unit	EDX-11A

Figure 1. Strain gauges connecting elements.

3 EXPERIMENTAL THEORY AND TOPOLOGY DESIGN

This study takes the structure of the solar heat storage-tank as an example. The structural material is stainless steel angle steel 304 (SUS304). The preliminary analysis results are obtained through the Abaqus software to simulate the static load of the structure. The horizontal and vertical seismic forces are designed according to the Ministry of the Interior's seismic design code. The results of the seismic forces on each axis were obtained by static analysis. Combined with time-series dynamic analysis, three-axis acceleration is applied to the structure to explore the structural deformation under real earthquakes. Finally, through topology optimization, the goal is to reduce the overall structure volume by 10% to obtain an optimized structure.

3.1 Experimental specifications

This research is based on the support structure of the solar thermal storage tank. First, SolidWorks is used to establish the model, and finally, the Abaqus software is used to calculate the force after the earthquake and obtain the topological support shape. Figure 2 is a structural assembly (excluding heat storage tanks, solar panels, and ground plates). Figure 3 is a diagram used in the Abaqus analysis structure (including heat storage tanks, solar panels, and ground plates). Table 3 describes the specifications of the support structure of the solar thermal storage tank.

Figure 2. A structural assembly.

3.2 Dynamic analysis method

3.2.1 Linear time-depend analysis
The adjustment coefficient of linear analysis is $I/(1.4\alpha\_y F\_u)$ but prevents buildings from exceeding the yield strength in case of moderate to small earthquakes. For areas generally close to faults, the adjustment coefficient should not be lower than $I/(4.2\alpha\_y)$, and for the Taipei Basin, the adjustment coefficient

Table 3. Analysis specification.

Name	The capacity of the heat storage tank (L)	Weight of heat storage barrel (empty) (kg)	Tripod weight (kg)	Weight of single solar panel (kg)
Specification	500 L	72 kg	50 kg	36 kg
Name	The longest distance of the tripod (mm)	The widest spread of the tripod (mm)	The height of the tripod (mm)	The thickness of the angle steel (mm)
Specification	2504 mm	4470 mm	1681 mm	1mm~2mm~3mm

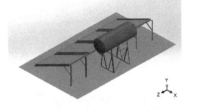

Figure 3. Analysis structure.

Table 4. Meinong earthquake data.

Earthquake direction	Vertical direction	North-south direction	East-west direction
Acceleration	132.26 gal	444.54 gal	426.22 gal

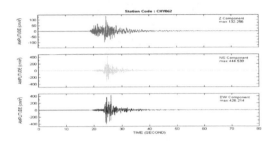

Figure 4. Meinong acceleration value.

should not be lower than I/(3.5α_y). The total transverse force in any significant axis direction obtained from the analysis adjusted following the specifications of the building's seismic design. When performing linear analysis, the damping ratio is higher after considering the interaction between the short-period structure and the soil. In the study, the equivalent damping ratio calculated according to the value of $S\_{\alpha}D/F\_u$ obtained from the static analysis. The equal damping ratio calculated according to the following formula (1):

$$\xi = \begin{cases} 5\% & ; S_{aD}/F_u \leq 0.3 \\ (16S_{aD}/F_u + 0.2)\% & ; 0.3 < S_{aD}/F_u < 0.8 \quad (1) \\ 13\% & ; S_{aD}/F_u \geq 0.8 \end{cases}$$

3.2.2 Nonlinear time-depend analysis

For nonlinear time-depend analysis, the structure's simulation carried out following the requirements of the building's seismic design. The nonlinear analysis model of the form must also accurately reflect the proper nonlinear behavior of the system. The response value obtained by nonlinear diachronic analysis did not reduce by the adjustment coefficient I/(1.4α_y F_u).

The dynamic no-linear analysis based on the 2016/02/26 Meinong earthquake test, the Richter scale is 6.6, and the maximum magnitude −7.0 (Tainan-Xinhua) test data uses Tainan City-Yujing District-Yujing Elementary School (CHY062). The seismic wave is shown in Figure 4. This analysis captures seismic waves from 15 seconds to 40 seconds and inputs the acceleration value every 0.005 seconds— the parameters obtained by analyzing more than 5000 pieces of data. The maximum acceleration in the triaxial direction of the Meinong earthquake is shown in Table 4.

In this study, we selected the site shear wave velocity in Huwei Town, Yunlin County, based on the geological data table of the station, and confirmed that the Huwei

site belongs to the second type of site (ordinary site); the F_α and F_v coefficients queried according to the table.

Site amplification factor of short-period structure:

(1) $S_{DS} = F_{\alpha}(S_S^D) = 0.7$ (2)

(2) $S_{MS} = F_{\alpha}(S_S^M) = 0.9$

Site amplification factor F_v for short-period structure:

(1) $S_{D1} = F_v(S_1^D) = 0.52$ (3)

(2) $S_{M1} = F_v(S_1^M) = 0.65$

4 TOPOLOGICAL STRUCTURE DESIGN

This thesis's primary purpose is to optimize the structure to achieve volume reduction and cost reduction without affecting the rigidity and strength. Topological optimization theory is based on design response strain energy and volume optimization methods to achieve structural volume reduction and rigidity maximization. This is done to optimize the topology to reduce the volume by 10% and design the M-shaped tripod for optimization. This study is based on the

Figure 5. Original structure frame.

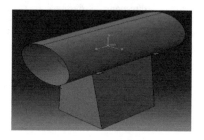

Figure 6. One-piece structure frame.

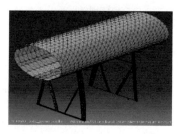

Figure 7. Original structure topology.

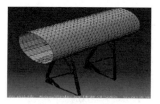

Figure 8. One-piece structure topology.

one-piece sheet structure applying lateral force in the 3-axial direction to solve the 3-axial force topological results. Finally, the original structure design and topology structure integrated into the new structure design. The initial structural design and the final topological structure design are shown in Figures 5–6. The optimized new structure design after the topology is shown in Figures 7–8. Figure 9 shows the final topology result.

Figure 9. Experimental equipment design.

Figure 10. Experimental equipment installation diagram.

5 EXPERIMENTAL EQUIPMENT DESIGN

This thesis's experimental verification method is to measure the magnitude of the forced strain with strain gauges and confirm whether the simulation analysis results are consistent with the actual measurement. Discuss the strain change of the original structure and topological structure under general static load and measure the effect. The equipment set-up is shown in Figure 10.

6 SIMULATION ANALYSIS AND RESULTS

6.1 *Hydrostatic pressure analysis*

This study's structural model was imported into Abaqus for static seismic simulation analysis and dynamic chronological simulation analysis.

According to the seismic force specification and other parameter settings, observe the structure's stress and displacement under the acceleration of 130 gals, 225 gals, 390 gals, and 444 gals in the three directions of the XYZ axis. Design the horizontal and vertical seismic forces according to the earthquake resistance laws and regulations. Multiply the coefficients by 9.8 m/s^2 and 100 to convert back to the seismic scale classification unit.

This research's experimental method is to fill the heat storage tank with water and measure the strain of the original structure and the topological structure with strain gauges. And verify whether the strain value is close to the analysis result. The results of the initial structure analysis showed in Figures 12–13. The maximum stress value is 77.56 MPa. Maximum strain value: $5.211 \times 10^{(-4)}$ The topological structure analysis result is shown in Figure 16. Maximum stress value:

Figure 11. The max. stress of the original structure.

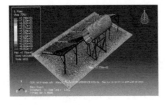

Figure 12. The max. strain of the original structure.

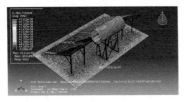

Figure 13. Topological maxi stress.

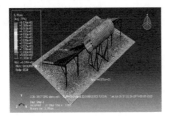

Figure 14. Topological max strain.

60.76 Mpa. Maximum strain value: $4.698 \times 10^{(-4)}$, as shown in Figures 14–15.

6.2 Strong seismic force analysis

According to the Central Meteorological Bureau's latest seismic acceleration class comparison table, the acceleration of the Level-6 strong earthquakes is 440~800 gal, and the coefficient is 0.45W. W is the weight of the water storage thermal bucket 4900N. The horizontal seismic force is 2205N, and the vertical seismic force is 1102N, and the XYZ axis. The results of the seismic force analysis showed in Table 5. The work shows that the structure is 235.4 MPa under the acceleration of 444gal in the +Z axis direction. According to the stainless-steel test standard, this result does not reach the yield strength of the material (318MPa).

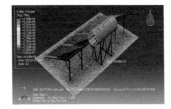

Figure 15. In +X direction stress.

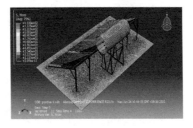

Figure 16. In +X direction strain.

Figure 17. In +Z direction stress.

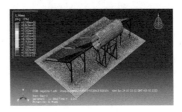

Figure 18. In +Z direction strain.

Table 5. Results of seismic force analysis with coefficient 0.45 W.

Seismic force 0.45W	Direction of force	stress	Displacement (mm)
Horizontal seismic force 2205N	+X	133.6 MPa	2.032×10^{-1}
	−X	82.28 MPa	1.590×10^{-1}
	+Z	235.4 MPa	5.345×10^{-1}
Vertical seismic force 1102 N	+Y	56.83 MPa	1.286×10^{-1}
	−Y	93.99 MPa	2.028×10^{-1}

91

7 CONCLUSION

This study analyzes the support structure through seismic acceleration simulation to obtain the topological optimization shape. Following the building's seismic design code, the force condition during a strong earthquake was obtained, and the dynamic time-depend analysis was used to simulate the real earthquake situation to get the structure's absolute safety. From the analysis results, it can be the view that the design in the Z direction is most affected by the seismic force, followed by the +X direction. Finally, try to reduce the structure's volume by 10% to achieve the same load as the original structure without damage. And can obtain the optimal topology design. Mainly summarized as the following three points:

1. According to the analysis of the building's seismic design specifications, the maximum acceleration (444gal) of the Mino earthquake, the horizontal seismic force coefficient is 0.45W the results show that the structure under the 444gal seismic acceleration reaches the maximum stress of 235.4MPa. According to the stainless-steel angle test standard provided by the manufacturer, this reaches the material yield strength.
2. In this paper, the optimal design of the supporting structure of the solar heat storage barrel is used to reduce the volume by 10%. The analysis results show that the top system has a maximum stress reduction of 16.8MPa and the maximum strain reduction of 0.513
3. Dynamic diachronic analysis, simulating the most realistic seismic forces affecting the design using the Mino earthquake to extract 15–40 seconds acceleration value for research. The results show that the maximum stress of the original system is 285.4 MPa. The maximum pressure of the topology structure is 289.5 MPa. Under the influence of proper triaxial acceleration, the two designs did not exceed the material yield strength. According to the above three points, the solar heat storage barrel's support structure can withstand earthquake acceleration below 444 gals. The main influence direction is the Z-axis. It recommended that the M-type tripod structure in the Z-axis direction be reinforced, and the final optimization result will reduce the volume by 10% This saves cost and weight for manufacturers.

REFERENCES

Hong Lifen, 2014, "Fatigue Analysis and Optimal Design of Universal Cylindrical CNC Grinding Machine," National Huwei University of Science and Technology Institute of Mechanical Design, Master's thesis.

Zhang Rongci, 2008, "Bridge design combining topology, size and shape optimization," Department of Civil Engineering, National Taiwan University, Master's thesis.

Yang Jiawei, 2019, "Earthquake-resistant and wind-resistant design evaluation and topology optimization of horizontal water towers," National Huwei University of Science and Technology, Master's Program of Mechanical Design Engineering, Master's Thesis.

Chen Yihong, 2019, "Vertical Water Tower Seismic and Wind Resistance Performance and Optimal Design," National Huwei University of Science and Technology, Master's Program of Mechanical Design Engineering, Master's Thesis.

Lin Liwei, 2012, "Analysis of the seismic resistance of steel building structures," Institute of Civil Engineering, Tamkang University, Master's thesis.

He Estekanchi, et al., 2007, "Application of Endurance Time method in linear seismic analysis," ScienceDirect-Engineering Structures, 29, 10, pp.2551–2562, October.

Hong Fan, et al., 2009, "Seismic analysis of the world's tallest building," ScienceDirect-Journal of Constructional Steel Research, 65, 5, pp. 1206–1215, May.

Case earthquake report, https://scweb.cwb.gov.tw/ special/special-index.htm

Smart Design, Science and Technology – Lam et al (eds)
© 2021 the Author(s), ISBN 978-1-032-01993-2

The dilemma of the sustainability of the fast fashion industry

Tien-Feng Hsu*
Ph.D Program of Mechanical and Energy Engineering, Kun Shan University, Tainan, Taiwan

Huann-Ming Chou
Department of Mechanical Engineering, Kun Shan University, Tainan, Taiwan

ABSTRACT: The Fast Fashion industry had swept across the globe since 2000. It had rapidly become a viable model of commercial success. However, in the past few years, the society gradually realizes the seriousness of contamination as well as abusive labor issues associated with the Fast Fashion garment industry. As a result, the general public demands that the industry ought to conform to sustainable development and to provide an improved labor environment. These put pressure on the participants of the Fast Fashion industry and since then, quite a few industry players had made declarations of sustainable development and advocated procedures to improve the working environment. However, some critics were still not impressed and commented that these efforts were not sufficient nor effective. They even consider this as part of a superficial campaign to "greenwash" the label of being a contaminator. This article hopes to investigate the dilemma surrounding the sustainability of Fast Fashion industry from the perspective of its operating strategies. This paper also includes feasibility suggestions leading to the industry's sustainable development.

Keywords: Fast fashion, Sustainable development, Circular economy, Renewable energy, Carbon emissions.

1   INTRODUCTION

1.1   *Research background and motivation*

Since about 2000, the success of brands such as Zara, H&M and Uniqlo has created a whirlwind of "fast fashion" around the world. In the Brand Finance Apparel 50 2019 report, released by Brand Finance (a brand valuation consultancy), Zara, H&M and Uniqlo occupied the 2nd, 4th and 7th places in the rankings respectively, while athletic brands Nike and Adidas ranked 1st and 3rd. This shows that fast fashion brands have become the leaders in the apparel industry. However, behind the success of fast fashion brands, there are serious problems with f pollution and labor exploitation. As the environmental problems caused by global warming are getting more attention, the rise in environmental awareness has brought along with its growth criticism of pollution problems created by fast fashion. Moreover, consumers are also becoming more receptive to products that are slightly more expensive, but more environmentally friendly. Meanwhile, the secondhand clothing market is also expanding, as concepts such as sustainable development and the circular economy become more widely accepted. As a result, fashion brands have also begun to think about how to meet sustainability requirements. For example, beginning in

2020, designers participating in Copenhagen Fashion Week must meet certain environmental requirements; while Inditex, the parent company of the fast fashion brand Zara, has also announced that it is aiming to use 100% sustainable materials and 80% renewable energy in all of its products by 2025. H&M, on the other hand, has announced a goal of 100% use of recycled or other sustainable materials by 2030; this goal has been 57% accomplished, according to the H&M Sustainability Report 2019. Nevertheless, many have seen this as a business tactic that will ultimately fail to achieve the Sustainable Development Goals (SDGs); other views are more optimistic, stating that citizens' environmental awareness will eventually bring fast fashion in line with sustainable development. The motivation for this study was to explore this issue.

1.2   *Research objectives*

The purpose of this paper was to investigate whether the fast fashion industry can achieve the SDGs, primarily through analyzing the business strategies of the fast fashion industry and understanding its core strategies and goals, and then also examining whether these strategies and goals are related to achieving sustainable development. Therefore, this paper will first explore the pollution and labor exploitation problems caused by the fast fashion industry, then analyze the

---

*Corresponding Author

DOI 10.1201/9781003188513-22

93

industry's core strategies and goals, and finally examine whether there is any conflict between the two. The aim is to examine whether the SDGs announced by the fast fashion industry are real goals that the companies embrace, or are simply a way to clean up the environmental mess that fast fashion has made.

2 THE PROBLEMS OF POLLUTION AND LABOR EXPLOITATION CAUSED BY FAST FASHION

2.1 Pollution led by fast fashion

The textile industry is second only to the petrochemical industry as a highly polluting industry. Take cotton as an example. It is the most commonly used item in the textile industry, yet it is estimated that only 2.4% of the world's farmland grows cotton; and growing cotton consumes 10% of agrochemicals and 25% of pesticides (Lu 2017). States that it causes great harm to land and water resources; it takes about 20,000 liters of water to produce 1 kilogram of cotton, and an estimated 8,000 chemical raw materials are used in the production of garments (Chen 2017). Claimed that in addition to consuming a great deal of resources, the discharge of wastewater also leads to serious pollution problems.

In addition to cotton, polyester and nylon are also commonly used as raw materials in the textile industry, and the manufacturing of these petrochemicals produces massive carbon emissions. Additionally, washing clothes made of these materials causes the microfibers in them to fall off and enter water systems such as sewers, rivers and oceans, and eventually enter living organisms and even the human bodies.

Also, in order to reduce costs, multinational companies usually locate their production bases in developing countries. At one time, this was primarily China, but due to the rising labor costs in China, multinationals have gradually shifted to Vietnam, Thailand, the Philippines, Cambodia, Bangladesh, Pakistan, etc. Raw materials are transported from the place of production to the factory, and the finished products are then distributed to sales outlets around the world. This process results in extremely high carbon emissions and air pollution.

Fast fashion is only exacerbating these pollution issues. The most characteristic feature of fast fashion is that it speeds up the design, production and sales of products, further increasing the speed of product turnover, which naturally increases the level of pollution. At the same time, the speed at which new products hit the shelves, along with their lower prices, can easily cause consumers to overspend, resulting in many discarded clothing. According to a survey completed by Greenpeace in January 2016 on shopping habits, Taiwanese people threw away 9.9 pieces of clothing every minute, or at least 5.2 million pieces of clothing per year, generating a lot of unnecessary waste of resources and waste disposal costs.

2.2 Labor exploitation led by fast fashion

On April 24, 2013, the Rana Plaza building collapsed in Dhaka, Bangladesh. The incident killed more than 1,000 people and injured more than 2,000, mostly women and children working in garment factories. The incident shocked the international community, and April 24 was designated as the "Fashion Revolution Day" one year after the incident in 2014, in the hope that the sweatshop phenomenon in the textile industry could be improved in the future.

According to an estimation drawn by the International Labor Organization in 2016, there were approximately 152 million child laborers between the ages of 5–17 worldwide, with nearly half of them, about 73 million, engaged in hazardous work. Whether adults or children, these textile workers earn extremely low wages, with workers in Brazil working 14 hours a day and receiving only €20 to 200 a month (Chen 2017). Found that the daily wage in a Cambodian garment factory was only US$3 per day for 18-hour work. Despite the low income, they have to work in factories full of toxic chemicals; but they still have to accept such harsh working conditions in order to make a living.

3 BUSINESS STRATEGIES AND GOALS OF FAST FASHION

The keyword for fast fashion is "fast": Faster design, faster mass production, and faster sales in the global stores. The product development cycle can be as short as 14 days. The aim is to present the trendiest products to consumers earlier than the competitors do, and to further stimulate consumers' urge to buy with the constantly updated products and relatively low prices. However, the quality of the clothing is actually sacrificed in order to lower selling prices. Nonetheless, it becomes easier for consumers to decide whether to buy or discard a piece of clothing without compunction, as a result of the lower prices and general quality of clothes available. In turn, this has even caused some internet celebrities to wear a shirt only once.

With fast speeds and affordable prices to accelerate product turnover rate and drive up the sales, while keeping purchase prices as low as possible and maintain gross margins, this is how the fast fashion makes profits. Table 1 is a summary of the five-year financial summary of Inditex, the parent company of Zara, showing that each year the business turnover has increased, with the gross margin has stayed above 55%.

4 THE DILEMMA OF SUSTAINABILITY IN THE FAST FASHION INDUSTRY

In recent years, environmental awareness has risen, which has also led fast fashion to become a target of public criticism. The high turnover rate and short life cycle of fast fashion products have further worsened pollution, so companies such as Zara, H&M,

Table 1. Five-year Inditex financial summary.

	2019	2018	2017	2016	2015
Net Sales	28,286	26,145	25,336	23,311	20,900
Gross Profit	15,806	14,816	14,260	13,279	12,089
Gross Margin	55.9%	56.7%	56.3%	57.0%	57.8%
Net Profit	3,647	3,448	3,372	3,161	2,882

Amounts in millions of euros

etc. have proposed sustainable development policies. However, every business runs to make a profit, and social responsibility is based on this premise. If a company loses money for an extended period of time and cannot continue operations, there will be no way to talk about social responsibility and sustainable development. On the other hand, the larger the company, the greater its impact on society, and the public and government will inevitably make certain demands on the company. Consequently, a company has to expend its own resources to meet these requirements, thus reducing its profitability; and so, how to strike a balance between the two becomes the crux of the problem. From past experiences, professional managers have tended to focus on maximizing profits at the expense of non-mandatory environmental and labor rights requirements, in order to be accountable to shareholders and maintain their positions. This is where the difficulty lies in moving a business towards sustainability.

5 CONCLUSION

From the above discussion, we can see that making a profit is ultimately the main goal of a company. It is not realistically enough to expect companies will fully make their own efforts towards sustainable development. We therefore suggest two areas for improvement:

1. Strengthening social education. Firstly, consumers should be educated to accept products that are slightly more expensive, but which are sustainable. Secondly, investors should be educated to support companies that are sustainable, not just profit-driven.
2. Government regulations must be kept up to date, and companies must be monitored for compliance with environmental and labor standards.

REFERENCES

W.Y. Chen, 2017a. The Ugly Truth Behind that Glamorous Display Window: The 'Fashion Pollution' You Need to Know About, The News Lens.
W.Y. Chen, 2017b. The Truth of Sweatshops Behind the Fashion Aura: Child Laborers Working 18 Hours for US$3.00 in a Garment Factory.
H.P. Lu, 2017. More Clothes than You Can Wear? The Secret of Fast Fashion's Effects on Environmental Pollution, Global Views Monthly.

Smart Design, Science and Technology – Lam et al (eds)
© 2021 the Author(s), ISBN 978-1-032-01993-2

The optimization and discussion of the direction control accuracy of synthetic shuttlecocks

Huei-Yung Liu*
Ph.D Program of Mechanical and Energy Engineering, Kun Shan University, Tainan, Taiwan

Huann-Ming Chou
Department of Mechanical Engineering, Kun Shan University, Tainan, Taiwan

ABSTRACT: In all kinds of ball sports or competitions, in addition to the player's own familiarity with the sport and skillfulness and the quality of the ball hit, the direction control accuracy when the ball is hit is also one of the important factors contributing to whether the target, which is the ball operated in the process, can achieve the direction and location expected and set by the striker.

In this paper, synthetic shuttlecocks and feather shuttlecocks in badminton were targeted to compare the difference between the two. Among the many differences, this study focused on the direct link between the direction control of the shuttlecock and the time (recovery time) required for the ball to change the direction when hit.

In addition, through the analysis of the actual measurement data, the preferable weight distribution ratio of the head and body of a synthetic shuttlecock was deduced. Improvement methods and recommendations for synthetic shuttlecocks to better resemble feather shuttlecocks were proposed. The purpose is to improve the on-site performance of synthetic shuttlecocks and enhance player and market acceptance towards synthetic shuttlecocks.

Keywords: Badminton, Shuttlecock, Feather, Synthetic, Direction control, Recovery time.

1   INTRODUCTION

In this paper, we studied badminton that everyone is familiar with. However, in order to be approved by the related organization and accepted in official competitions, a shuttlecock is required to have qualified performance in terms of its corresponding characteristics, such as hitting feel, flight path, direction control accuracy, right speed and durability.

The direction control accuracy is the only characteristic to be discussed in this paper, and the focus will be placed on the comparison and investigation of the weight distribution between the feathered and synthetic shuttles, as it is the main factor that directly affects the direction control accuracy of a shuttlecock.

Among the different materials used in making synthetic shuttles in the industry, this study compared the nylon synthetic shuttles (Figure 1) with the feathered shuttles (Figure 2). The conclusion of the comparison between the theory and the actual measurement values will put forward reasonable suggestions and feasibility in regards to the improvement of the synthetic shuttle simulation.

Figure 1.   Nylon synthetic shuttles.

2   RESEARCH METHODOLOGY

2.1   Direction control accuracy

Badminton is a back-and-forth motion, pairing game in which the players must deliberately perform different strokes to move the racket with the desired direction and to the location. These strokes include forehand, backhand, high clear, clear, drive, smash, cut, net shot, etc. However, the degree to which the shuttlecock itself provides accurate direction control

---

*Corresponding Author

DOI 10.1201/9781003188513-23

Figure 2. Feathered shuttles.

Figure 4. Cork base.
Source: Keyluck Ind. Corp. (Shuttlecock Manufacturer)

is an immediate and important factor in achieving the desired result.

The direction control accuracy of shuttlecocks is closely related to its recovery time while being hit; the shorter the recovery time (i.e. speedy), the more accurate the direction control of a shuttlecock. Therefore, the excellent performance of recovery time has always been one of the key points in the research and development of shuttlecocks.

2.2 Recovery time

The recovery time, in other words, the flying trajectory of a shuttlecock is shown in Figure 3. This figure cites an experiment conducted by Cookke, A.J. (Texier et al. 2012) as an example, in which high-speed photography of recordings was used to analyze the recovery time, which the industry also has it known as the "Speedy Turnover" object, i.e. the time required to change the flight path of a shuttlecock when it is hit, from its instantaneous turnover to fully stabilized for the reverse flow.

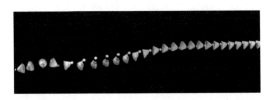

Figure 3. The recovery flying trajectory of a shuttlecock being hit, using the high-speed photography method. (The racket is moving from left to right; the photo was taken with the shutter speed at 0.005 seconds per frame [5ms]).

For centuries, the materials used to make the feathered shuttles, such as cork base (Figure 4) and goose or duck feather, have been set and it would be difficult to change it drastically. Nevertheless, after years of research and development efforts made by the manufacturers, the characteristics performance of a nylon synthetic shuttle has become more similar to that of a feathered shuttle

Figure 5 shows the recovery process of a nylon synthetic shuttle taken by using high-speed photography method. The figure shows the recovery image and process of the nylon synthetic shuttle, while the statement is retrieved from the promotional description of the nylon synthetic shuttle by Yonex, model Mavis-370. According to the figure, the recovery time of Yonex's feathered shuttle is 0.015 seconds, while its nylon synthetic shuttle (Mavis-370) being 0.020 seconds, with only a 0.005 seconds difference exists wherein (Catalog of Yonex, Nylon shuttles, Product Technology, Recovery time). In addition, the figure indicates that the recovery performance of a nylon synthetic shuttle at this stage is getting closer to that of a feathered shuttle. Objectively speaking, many of the nylon synthetic shuttles manufactured by the other brands have the similar excellent performance

When smashed, a MAVIS shuttlecock recovers in only 0.02 seconds. This performance is just 0.005 seconds slower than a YONEX Feather shuttlecock and 0.008 seconds faster than the recovery of an ordinary shuttlecock.

Figure 5. The recovery process taken by using the high-speed photography method.

2.3 Weight ratio between the front and rear of a shuttlecock

The cork base at the front of a shuttlecock is much denser and heavier than the fluffy skirt at its rear. Thus, when a shuttlecock is hit, the heavier front end will lead the lighter rear end to flow reversely. Nonetheless, if the weight between the two ends is not significant, the shuttlecock will recover slower and less neatly, resulting in a clumsy wobble and longer time to complete the recovery process. Therefore, the weight between the front and rear of a shuttlecock has a decisive influence on its recovery time.

2.4 Weight limits under the badminton shuttlecock equipment approval certificate

The Badminton World Federation (BWF) has a clear definition of what a shuttlecock should look like together with its rules; meanwhile it is used as the basis for the approval standards of a shuttlecock.

In this section, only the rules relating to the weight of the shuttlecocks discussed in this study are listed here:

1. Feathered Shuttle
   The shuttlecock shall weigh between 4.74 to 5.50 grams.
2. Non-Feathered Shuttle
   Measurements and weight of the shuttlecock shall be as in Rules 2-2, 2-3, and 2-6. However, because of the difference in the specific gravity and other properties of synthetic materials in comparison with feathers, a variation of up to 10% shall be accepted. (Badminton World Federation (BWF) 2019, May)

Hence, the weight adjustment of a synthetic shuttle should be limited to between 4.27 and 6.05 grams, in order to meet the rules. This is also one of the necessary considerations for manufacturers in the research and development of shuttlecocks.

3 ACTUAL MEASUREMENT AND DISCUSSION ON THE WEIGHT OF THE CORK BASE AND SKIRT IN SHUTTLECOCKS

In this study, the tournament grade feathered shuttles and the nylon synthetic shuttles of two famous brands (Victor from Taiwan and Yonex from Japan) were compared and analyzed.

First of all, we took one sample from each of the four types of Victor and Yonex feathered and nylon synthetic shuttles; their front cork base and the rear skirt were cut and disassembled (Figure 6), and were weighed respectively. Then, the ratio of the two parts in each shuttlecock was then calculated, as shown in Figures 7–10:

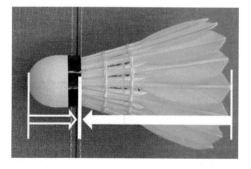

Figure 6. The front cork base and the rear skirt were cut and disassembled.

Figure 7. Weight measurement of victor B-01 feathered shuttle.

Figure 8. Weight measurement of Yonex AS-50 feathered shuttle.

Figure 9. Weight measurement of victor NS-3000 Nylon Synthetic shuttle.

Figure 10. Weight measurement of Yonex Mavis 370 Synthetic shuttle.

Based on the results of the four types of shuttles above (Figures 7–10), the weights of feathered and nylon synthetic shuttles between their front and rear ends were summarized, as shown in Table 1.

From the data presented in Table 1, the weight ratios between the feathered and nylon synthetic shuttles are described as follows:

1. The average weight ratio between the front cork base and the rear skirt in the feathered shuttles is 1.34:1.

Table 1. Weight ratio of the shuttlecocks.

Shuttlecock Types	Total Weight	The Cork Base at the Front	The Skirt at the Rear	Front and Rear Weight Ratio	The Average Weight Ratio
Victor BN-01 Natural	5.17	2.97	2.20	1.35:1	1.34:1
Yonex AS-50 Natural	5.09	2.91	2.18	1.33:1	
Victor NS-3000 Nylon	5.44	2.86	2.58	1.11:1	1.13:1
YY Mavis370 Nylon	5.13	2.74	2.39	1.15:1	

2. The average weight ratio between the front cork base and the rear skirt in the nylon synthetic shuttles is 1.13:1.

From the table, it can be seen that the weight ratio between the front and rear ends in feathered shuttles is obviously higher than that of the synthetic shuttles. That is to say, the feathered shuttles require shorter recovery time than the nylon synthetic shuttles when being hit. If the recovery time is the only variable changing, holding all other variables constant, the direction control accuracy of a feathered shuttle will be superior to the performance of a nylon synthetic shuttle. From the perspective of badminton players, the shuttles with higher direction control accuracy will be perceived as the "more obedient" ones.

Since the two feathered shuttles sampled in this study are both internationally recognized tournament grade shuttlecocks, they have been accepted, accustomed and recognized by players for a long time in terms of their hitting feel, direction control accuracy and other characteristics. Therefore, the weight ratio of the feathered shuttle (1.34:1), which is based on the abovementioned actual measurement values, can be reasonably regarded as the golden ratio that manufacturers are seeking in the research and development of synthetic shuttles.

4 CONCLUSION

In recent years, the shuttlecock manufacturers and the industry have been improving the structure and material of shuttlecocks, in the hope to develop a type of synthetic shuttle that is as feather like or close to feather as the tournament grade feathered shuttle does. Therefore, if the weight ratio of the shuttles can be as close as that of the feathered shuttles (1.34:1), along with other improvements, we can optimistically expect that the synthetic shuttles will gradually replace the feathered shuttles partially, or even one day completely, as well as be used in official competitions internationally.

REFERENCES

Catalog of Yonex, Nylon shuttles, Product Technology, Recovery time.
Baptiste Darbois Texier et al, 2012. Procedia Engineering, 34, pp. 176–181.

Smart Design, Science and Technology – Lam et al (eds)
© 2021 the Author(s), ISBN 978-1-032-01993-2

The composition properties of wood vinegar produced from pyrolyzed palm kernel shell waste

Chien-Yuan Chen*
Ph.D Program of Mechanical and Energy Engineering, Kun Shan University, Tainan, Taiwan

Hsi-Shou Lee
General Manager, Yuan Da Technology Co., Ltd.

Huann-Ming Chou
Department of Mechanical Engineering, Kun Shan University, Tainan, Taiwan

ABSTRACT: In this study, an anoxic pyrolysis mass production process, incorporating the application of the near-slow pyrolysis method, was used to decompose palm kernel shell waste at 500-550° to produce biochar, biogas, and liquid wood tar mixture products. The liquid wood tar mixture was stood for phase separation, then distilled to produce heavy metal-free wood vinegar with a pH value of about 3.5~4.1 and organic matter content of about 5.5%, of which the nitrogenous organic matter content was 0.2%. Further GC/MS analyses revealed its main component to be acetic acid and the presence of at least 32 kinds of organic substances, such as phenols, nitrogen-containing compounds, etc. Throughout the storage period, the color changed from light yellow to brown and the organic components also changes via oxidation or reduction reaction.

Keywords: Palm kernel shell waste, Pyrolysis, Bio-oil, Wood vinegar.

1   INTRODUCTION

Pyroligneous acid, also known as wood vinegar, is a chemical generated from the decomposition of biomass or lignocellulosic waste at a high temperature under low oxygen or anoxic conditions; an aqueous liquid, produced from the condensation of the steam and smoke during the process, is called bio-oil or wood tar. When placed in a closed container, the liquid separates into two layers, with the lower layer being brown or dark brown tar sediment and the transparent liquid in the upper layer being crude wood vinegar. The vinegar gives off a special smoky odor and has a color between light yellow and brown; the color varies depending on the properties of the raw materials and the operating conditions of the pyrolysis system used for its preparation. The substance obtained through processing of the crude wood vinegar with methods such as filtration, distillation, or extraction, is referred to as wood vinegar or pure wood vinegar and has a higher application value.

The major constituent of wood vinegar is water (80–90%), which contains more than 200 organic compounds. Recently, wood vinegar has been widely used in various applications, such as smoky aromas,

food, platelet aggregation and medicine with anti-fungal activity targeting skin fungi (Amen-Chen et al. 2001; Loo et al. 2007). In organic agriculture, natural wood vinegar has replaced many toxic chemicals used for disease and pest defense, plant growth stimulation, fruit quality improvement, and plant seed growth and germination acceleration, and as an herbicide (Yoshimoto 1994). However, the composition, physicochemical and biological activities of wood vinegar are affected by many factors, including the chemical composition of the biomass, pyrolysis system, and purification methods. The influence of the operating conditions used in the pyrolysis process on the production rate of the products, such as biochar, wood tar, and biogas (Bridgwater et al. 1999; Yang et al. 2006), the optimal operating conditions for wood tar production (Wada 1997), and the instability of wood vinegar composition have been discussed in literature (Wada 1997). Some of their physical properties, such as pH, density, color, odor, dissolved tar content, burning residues and transparency, have also been reported (Mun et al. 2007). However, the influence of storage time on changes of wood vinegar composition have not been extensively researched.

In this study, the crude wood vinegar produced by the pyrolyzing process of palm kernel shell waste in a newly constructed mass production plant (YD Com.) in Taiwan was investigated. The pure wood vinegar

*Corresponding Author

DOI 10.1201/9781003188513-24

Figure 1. Dried palm kernel shell waste.

Figure 2. The colors of wood vinegar samples stored for different time lengths (provided by Yuan Da Technology Co., Ltd.)

after distillation was stored for different time lengths, samples were acquired and the changes of its composition were analyzed with GC/MS to provide important information for applications.

2 EQUIPMENT AND METHODS

2.1 Materials and manufacturing process

The palm kernel shell waste (irregular granules, length and width between 5~10 mm), as shown in Figure 1, was imported from Indonesia; the moisture and oil content were 11.4 and 0.44%, respectively. The manufacturing units of the palm kernel shell waste pyrolysis factory included an anoxic cracking furnace (gasifier), a washer, a distillation tank, etc. The palm kernel shell waste was first sent to the biomass gasifier with a volume of around 1 m³ via a conveyor belt, and the temperature in the gasifier was increased to 500~550°. The palm kernel shell waste was gradually carbonized into biochar, and the biogas produced by smoldered cracking was sucked from gas pipes attached to the sides of the gasifier into two washing towers with volumes of around 55 and 1100 liters to produce a wood tar-water mixture (referred to as crude wood vinegar) through condensation, which was first stored in a storage tank, then subjected to a distillation process to obtain pure wood vinegar, while the distillation residue was wood tar. Wood vinegar was sampled and analyzed by GC/MS during the storage period to observe the changes of the organic compound constituents. Samples were collected directly (stored for 0 days), stored for 10, 30, or 180 days as shown in Figure 2 (from right

to left), indicating that the longer the storage time, the darker the color of the wood vinegar.

2.2 Analysis equipment

A GC-MS system (Varian, USA), comprised of a CP-3900 gas chromatograph (Walnut Creek, USA), a 1177 injector, a CP-8410 auto-sampler and an ion-trap mass spectrometer (Varian Saturn 2100T), was used to examine the samples. Two different stationary phases, the VF-5ms (Varian Factor Four) fused-silica capillary column (internal diameter 30 m × 0.25 mm, film thickness 0.25 mm) and the DB-WAX (J&W Scientific, USA) fused-silica capillary column (inner diameter 30 m × 0.25 mm, film thickness 0.25 mm), were used for the analysis. The oven temperature for the VF-5ms column was set as the following: 60°C for 1 minute, followed by heating to 280°C with a temperature increase rate of 6°C min-1 (then held for 3 minutes). The chromatography mass spectrograms of the samples are shown in Figure 3.

3 RESULTS AND DISCUSSION

Each ton of palm kernel shell waste could produce approximately 0.18~0.22 tons of wood tar-water mixture, which, after distillation, could further produce 0.15~0.20 tons of wood vinegar and 0.02~0.03 tons of wood tar. The organic matter content of the vinegar was about 2.5~6.5% (mainly acetic acid and phenolic components), and the pH value was around 3.1~4.2. The compositions obtained from qualitative analyses of the chromatography mass spectrum in Figure 3 is shown in Table 1, which revealed that the composition of wood vinegar was not entirely the same after different storage periods after distillation. When compared with day 0 of storage, there were 12 components that were the same on day 10 of storage (in bold), including acetic acid, acetoin, butyric acid, creosol, 2-cyclopenten-1-one, ethanol, furfural, 1-hydroxy-2-butanone, phosphorus pentafluoride, propanoic acid, 2-pyridiamine, and N-methyl-.

The number of the other different ingredients increased from 21 to 23 types. When comparing day 30 of storage to day 0, there are only 5 ingredients that were the same, including acetic acid, creoso, 2-cyclopenten-1-one, furfural and propanoic acid; the number of the other different ingredients increased from 21 to 25 types. When comparing day 180 of storage to day 0, the only same ingredients were acetic acid, creoso, 2-cyclopenten-1-one, and propanoic acid; the number of the other different ingredients decreased from 21 to 17. From this result, it could be inferred that decomposition or polymerization of wood vinegar occurred during storage, which was possibly caused by some free radicals or unstable compounds generated by the pyrolysis process. Therefore, the standardization of wood vinegar products is currently still difficult to be implemented.

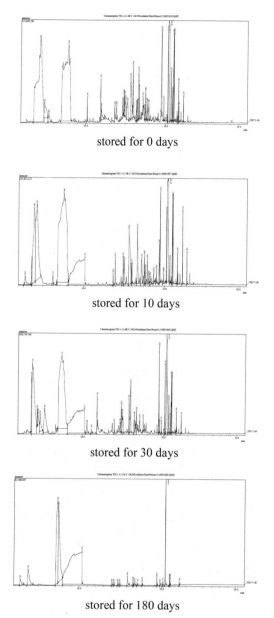

stored for 0 days

stored for 10 days

stored for 30 days

stored for 180 days

Figure 3. GC/MS chromatography mass spectrums of wood vinegar stored for different time lengths after distillation (Data provided by Yuan Da Technology Co., Ltd.)

4 CONCLUSION AND SUGGESTIONS

The palm kernel shell waste produced wood tar under anoxic pyrolysis, which could be distilled to produce wood vinegar. Water, acetic acid and phenol were the major components. It had a pH value of around 3.3-4.1, and its organic matter content was around 5.5%; the color, odor, dissolved tar content, and transparency have been evaluated using the quality index of wood vinegar. In this study, it has been discovered that the color and composition of wood

Table 1. The transformation of components in the wood vinegar analyzed by GC/MS after different days of storage (provided by Yuan Da Technology Co., Ltd.)

After 0 days of storage

*Acetic acid,BAcetoin,BButyric acid,BCreosol,B2-Cyclopenten-1-one,BEthanol,BFurfural,B1-Hydroxy-2-butanone,BPhosphorus pentafluoride,BPropanoic acid,B2-Pyridiamine, N-methyl-,B*Acetic anhydride,BAcetohydroxamic acid,BBenzene, 1,4-dimethoxy-2-methyl,BCyclopentanone,BCyclobutane, methyl-,B 2-Cyclopenten-1-one, 2- methyl-,BCyclopropane,B5-Ethyl-2-furaldehyde,BEthanone,1-(2-furanyl)-,BFuran<2-acetyl>,B2-Furanmethanol,B1-Hexen-3-yne,BIsoxazolidine,5-ethyl-2,4-dimethyl-, trans,B3-Methylpyridazine,BNitrous oxide,B1,4-Pentadiene, 3-Methyl-Phenol, 3-ethyl-,B2-Propen-1-ol,B2-Propanone, 1-hydroxy,BSyringol,BTetraborane(10) ,BXylenol<2,6->

After 10 days of storage

Acetic acid, dichloro, Acetoin, Butyric acid, Creosol, 2-Cyclopenten-1-one, Ethanol, Furfural, 1-Hydroxy-2-butanone, Phosphorus pentafluoride, Propanoic acid, 2-Pyridiamine, N-methyl-, Alpha-Aminoisobutyronitrile, Ammonia, Butyrolactone, BCyclopentan-1,2-dione<3,5- dimethyl ->, 2-Cyclopenten-1-one, 3- methyl-, 2-Cyclopenten-1-one, 3-ethyl-, 2-Cyclopenten-1-one, 3-ethy-2-hydroxy-, 3,4-Diacetylfurazan, Di(1,2,5-oxadiazolo)[3,4-b:3,4-E]pyrazine, 4,8, 5-Ethyl-2-furaldehyde, Formic acid anhydrazide, 2-Furanone, 2,5-dihydro-3,5-dimethyl, Guaiacol<4-ethyl->, 2(5H)- Furanone, 3- methyl-, 1-Hexen-3-yne, Meso-3,4-Hexanediol, 3-Methylcyclopentan-1,2- dione, Nitrogen, 3-penten-1-yne, Phenol, 2,6-dimethoxy-, 2-Propanone, 1-(acetyloxy)- , Pyridine, 3-methoxy-, Syringol<4-methyl->

After 30 days of storage

Acetic acid, Creosol, 2- Cyclopenten-1-one, Furfural, Propanoic acid, 3-Piperidinol, 1,4-dimethyl-, cis-, Bis(N-methoxy-N-methylamino)methane, Methyl Alcohol, 2-Butanone, 2-Propyn-1-ol, acetate, Methanethiol, Ammonia, 2-Propen-1-ol, 1-Butanol, Cyclopentanone, 2-Propanone, 1-hydroxy-, 2- Cyclopenten-1-one, 2-methyl-, 1-Hydroxy-2-butanone, Furan<2-acetyl>, Ethanone, 2-cyclopentyl-1-(1H-imidazol-4-yl)- , Butyric acid , 5-Ethyl-2-furaldehyde, 2-Fluorobenzoic acid, 4-nitrophenyl ester, Xylenol<2,6>, 3-Penten-1-yne, Phenol, 4-ethyl-2methoxy-, Phenol, 2-ethyl-, Phosphorous pentafluoride, Phenol, 2-methyl-, Syringol

After 180 days of storage

Acetic acid, Creosol, 2-Cyclopenten-1-one, Propanoic acid, Nitrogen, Acetone, Piperidine, 2-propyl-,(S)- , Thionyl chloride, Ammonia, 2-Propanone, 1-hydroxy-, Pyrazine, 2,6-dimetyl-, 1-Hydroxy-2-butanone, 2-Cyclopenten-1-one, 3-methyl-, Pyridine, 3-methoxy-, Butanoic acid, 4-hydroxy-, Phenol, 2-methoxy-, Phenol, 3-Methylpyridazine, p-Cresol, Phenol, 3-methyl-, Syringol

vinegar were affected by storage time lengths; the longer the time, the darker the color. Wood vinegar that has just been distilled had at least 35 organic components that could be detected by GC/MS, and decomposition or polymerization reactions possibly occurred and altered its compositions during storage (0~180 days). However, as a natural product, wood vinegar still has beneficial applications in agriculture, the pharmaceutical industry and biomedicine, veterinary and animal production, food processing and wood preservatives, etc.

ACKNOWLEDGMENTS

We are grateful to Yuan Da Technology Co., Ltd. for the providing of the information and data and to Dr. Jun-Ming Cheng for his guidance; without their help, the successful completion of this research would not have been possible.

REFERENCES

Amen-Chen C, Pakdel H, Roy C, 2001. Production of monomeric phenols by thermochemical conversion of biomass: A review, Bioresource Technology, 79:277–299.

Bridgwater AV, Meier D, Radlein D, 1999. An overview of fast pyrolysis of biomass, Organic Geochemistry, 30(12):1479–1493.

Loo AY, Jain K, Darah I, 2007. Antioxidant and radical scavenging activity of the pyroligneous acid from a mangrove plant, Rhizophora apiculate, Food Chemistry, 104:300–307.

Mun S, Ku C, Park S, 2007. Physicochemical characterization of pyrolyzates produced from carbonization of lignocelluloses biomass in a batch-type mechanical kiln, Journal of Industrial and Engineering Chemistry, 13:127–132.

Wada T, 1997. Charcoal Handbook, Forest Management Section, Agriculture, Forestry and Fisheries Division. Japan: Bureau of Labour and Economic Affairs, Tokyo Metropolitan Government.

Yang H, Yan R, Chen H, Zheng C, Lee DH, Liang DT, 2006. In-depth investigation of biomass pyrolysis based on three major components: Hemicellulose, cellulose and lignin. Energy & Fuels, 20:383–393.

Yoshimoto T, 1994. Present status of wood vinegar studies in Japan for agriculture usage, In:Proceeding of the 7th International Congress of the Society for the Advance of Breeding Researches in Asia and Oceania (SABRAO) and International Symposium of World Sustainable Agriculture Association, pp. 811–820.

Smart Design, Science and Technology – Lam et al (eds)
© 2021 the Author(s), ISBN 978-1-032-01993-2

Influence of printing temperature and speed on the mechanical properties of 3D-printed PETG parts

Ming-Hsien Hsueh*, Yi-Jing Su, Cheng-Feng Chung & Chia-Hsin Hsieh
Department of Industrial Engineering and Management, National Kaohsiung University of Science and Technology, Kaohsiung, Taiwan

Chao-Jung Lai, Jui-Fang Chang
Department of International Business, National Kaohsiung University of Science and Technology, Kaohsiung, Taiwan

ABSTRACT: The purpose of the research is to improve the precision and optimize the mechanical characteristics of 3D-printing technology, based on Fused Deposition Modeling (FDM), by adjusting the temperature and printing speed, in order to maximize the productivity and to minimize the material cost of Polyethylene Terephthalate Glycol (PETG). This research also analyzed the effects of temperature and printing speed parameters on these two materials, including their precision and material properties, and the following results were obtained: (1) Increasing the printing temperature in the material processing range improves the tensile strength, compressive strength and flexural strength of the material significantly. (2) In order to optimize the mechanical characteristics of precision and the heat-resistant ability of the PETG material, the printing parameter should be set at a processing speed of 30m/s, and the nozzle temperature should be 245°C, which could decrease the deviations of the product and obviously maximize the heat distortion. (3) When the mechanical characteristics are focused on the heat-resistant ability, the printing parameters of the processing speed and nozzle temperature should be set at 25m/s and 225°C, respectively.

1 INTRODUCTION

3D-printing, also called Advanced Manufacturing (AM), has attracted the attention of manufacturers with increased competitiveness. Compared with the manufacture of traditional tools, 3D printing has a lower processing time, and lower material and labor costs. 3D printing could match the higher complexity of the product. In the technology of 3D printing, Fused Deposition Modeling (FDM) has the lowest manufacturing cost. Therefore, much research that is based on FDM focuses on the influence of temperature and printing speed parameters on the mechanical properties. Roundy et al. (2002) [1] analyzed the effects of the compressive strength of the upright and horizontal direction of Acrylonitrile Butadiene Styrene (ABS). Ebel et al. (2014) [2] explored the effects of two different infill patterns on the mechanical properties of ABS and PLA, and the results showed that PLA has a higher Young's modulus than ABS. Badrossamay et al. (2015) [3] studied the effects of the gap distance on the mechanical characteristics of PLA. The results showed that 210°C is the best nozzle temperature, and

that it could minimize the deviations of the air chamber. Hernandez et al. (2014) [4] analyzed the material characteristics of PLA by adding Polyethylene Glycol (PEG), and the results showed that the surface wettability, biodegradability and roughness all increased, and that 5% (w/w) is the best adjunction ratio. Valerga et al. (2018) [5] analyzed the effects of the printing temperature, the humidity and the color of the wire on the mechanical properties, based on the PLA, and the results showed that a higher printing temperature will enhance the deviations of the product size and reduce the tensile strength. Jiang et al. (2017) [6] explored the tensile strength and the fiber distribution of PLA, ABS, PETG (Polyethylene Terephthalate Glycol) and Amphora, by adding corban fiber, and found that the raster angles of printing are 0°, 45°, ±45° and 90°, respectively. Santana et al. (2018) [7] explored the different mechanical properties of PLA and PETG, based on FDM and injection molding technology. The present paper on PETG shows that it has a higher heat resistant ability than PLA; therefore, many producers are expected to use PETG to make up for the disadvantages of PLA. This research will focus on analyzing two parameters, the nozzle temperature and the process speed, to explore their effects on the mechanical behavior of PETG.

---

*Corresponding Author

DOI 10.1201/9781003188513-25

Table 1. The specifications of the 3D printer.

Name	Specification
Physical dimension	(w)40cm × (d)22cm × (h)46cm
Maximum printing area	(w)20cm × (d)20cm × (h)24cm
Print layer height	0.04~0.32 mm
Wire diameter	Φ1.75mm
Nozzle diameter	0.2, 0.4, 0.6 mm
Platform temperature	~110°C
Nozzle printing temperature	~300°C
Cooling method	4.5 cm turbo fan

Table 2. The specifications of the PETG.

Name	Specification
Color	Matt black
Wire diameter	1.75 ± 0.05 mm
Weight	800 g
Recommended printing temp	215~235°C
Recommended printing speed	30~50 mm/s

Table 3. Process parameter conditions of the printing material for KISSlicer software.

Content Project	Name	Range
Controlling factor	printing speed	25~30 mm/s
	Nozzle printing temperature	225~235°C
Fixed factor	Fill rate	20 %
	Layer thickness	0.2 mm

Table 4. The specifications of the QC-H51A2 universal testing machine.

Type	H51A2
Capacity	100 kN
Stroke	1100 mm(without fixture)
Space	Ø550 mm
Load Resolution	1/10,000 (maximize 1/200,000)
Displacement resolution	0.001 mm
speed	0.003~375 mm/min
height	2,200 mm
Current	15 A

2 EXPERIMENTAL

This research analyzes two specific parameters, namely, the processing speed and nozzle temperature that affect the mechanical behavior of the materials of PETG. The materials were used to print standard test specimens, such as ASTM D638, ASTM D648, ASTM D790, and ASTM D3410, in order to evaluate their tensile strength, compressive strength, bending strength, and heat resistance. The software KISSslicer was used to adjust the printing parameters of the nozzle temperature and process speed, respectively, by fixing the other parameters and by analyzing the difference between the 30 mm/s and 25 mm/s processing speeds of the PETG specimens. The results could forecast the trends in the characteristics of the material after changing the printing parameters.

The 3D printer used in this research is the X1E produced by Yuanli Think Tank Co., Ltd. (Kaohsiung, Taiwan), and the specifications of printer are shown in Table 1. The PETG materials that were used in this study were supplied by Min-Yau Information Co., Ltd (New Taipei, Taiwan), and the specifications of the printing materials are shown in Table 2.

SolidWorks software was used to build the 3D model of the test specimen. The slicing software used in this research is KISSslicer, which provides multiple parameter settings. The specifications of the software are shown in Table 3. Based on the highest efficiency, this research set the process speed in a specific range to avoid the cave structure caused by incomplete cooling, due to the high speed. In the literature mentioned above, the higher infill percentage caused the higher firm structure, but it took more time to process. Therefore, the percentage of the infill pattern was set as a 20% rectilinear structure, the layer thickness was set at 0.2mm and the size of the selected nozzle was 0.4 mm, to maximize the efficiency and to minimize the loss of material. The specifications and settings of the software printing parameters are shown in Table 3.

To evaluate the tensile, compression, and bending strengths, the QC-H51A2 universal testing machine was used. The specifications of the testing machine are shown in Table 4. The data strength of the test, with a force of 100kN, are recorded by the built-in program. The data collected from the test were analyzed by using Excel software to calculate the values of Young's modulus, as well as the ultimate strength, and the yield strength (0.2% offset) of the specimens. The tensile specimens were prepared according to ASTM D638 standard testing, the compression specimens were prepared according to ASTM D3410 standard testing, and the bending specimens were prepared according to ASTM D790 standard testing.

3 RESULTS AND DISCUSSION

3.1 Tensile testing

The results of the tensile testing of PETG are depicted in Figure 1. Figures 1(a) and (b) show the results of the tensile testing of the specimens processed in 30mm/s and 25mm/s, respectively. The specimens processed at each temperature are demonstrated in Figure 2. From the stress strain curve exhibited in Figure 1, it can be seen that the mechanical properties of PETG have the same trend with the brittle material. When the nozzle temperature is over 235°, there is an obvious

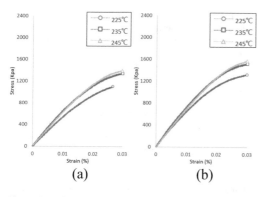

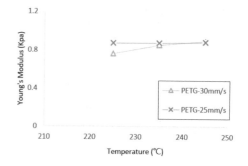

Figure 1. The stress-strain curve for the tensile testing of PETG (a) 30mm/s, (b) 25mm/s.

Figure 3. Young's modulus for the tensile testing of PETG specimens.

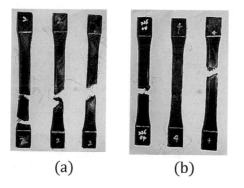

(a)   (b)

Figure 2. Specimens of the tensile strength of PETG printed in each temperature (a) 30mm/s, (b) 25mm/s.

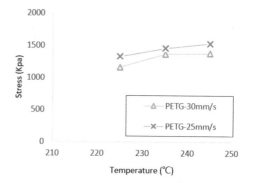

Figure 4. The UTS value for the tensile testing of PETG specimens.

increase in the tensile strength, Young's modulus and the average value of UTS, namely, 1374.71N, 0.87Kpa, and 1017.79Kpa, respectively. Compared with the obtained values of the specimens which printed at a temperature of 225°, the tensile strength increased by 18%, Young's modulus increased by 14%, and the average value of UTS increased by 26% of the specimens printed at a temperature of 235°. It is evident that Young's modulus was not enhanced significantly from 225°~245°. The results show that the effect of the nozzle temperature on Young's modulus is not obvious; instead, the value of the UTS and yield strength increased, as the temperature increased.

Figure 3 illustrates the relevance of Young's modulus to the nozzle temperature of PETG materials. Because of the high value of the condensation temperature of PETG, the results show that the matrix of material arrays closely at a low process speed. It also shows that there is an evident difference between the exterior of the specimens processed in 30mm/s and 25mm/s, when the temperature is over 225°, but there is apparently no difference between Young's modulus, which may be caused by having insufficient variations. The results indicate that the printing parameters of the process speed and nozzle temperature have no obvious effect on Young's modulus of PETG.

The value of UTS on PETG is demonstrated in Figure 4. The nozzle temperature must be over 190°C to maximize the value of UTS. The value of the PETG specimens (35mm/s) reduces when the temperature is over 190°C, due to the cave structure between the layers that results from the friction between the material and inner structure of the nozzle because of the slow process speed [5]. The results show that the yield strength and Young's modulus decreased because 5 mm/s of the reduction could still not array the materials closely. In addition, the results show that the trend of the PETG specimens is similar between the process speeds of 25mm/s and 30mm/s, and that the higher temperature causes the higher value of UTS. The obtained value indicates that PETG performed well on the value of UTS because of the slower speed of crystallization [7]. Take, for instance, the specimen processed in 25mm/s; there did not appear to be friction in the inner structure of the nozzle. The results show that when the mechanical characteristic focus on UTS, the process speed of PETG should be set at 25mm/s.

The value of the yield strength of PETG is demonstrated in Figure 5. The yield strength of 30mm/s and 25mm/s enhances with the temperature; however, the 25mm/s specimens have a higher value as a result of the lower crystallinity, the lesser cave structure, and the higher stickiness.

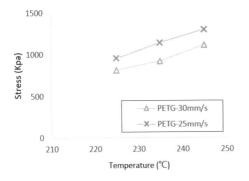

Figure 5. The yield strength for the tensile testing of PETG specimens.

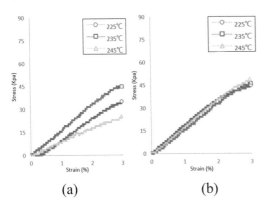

(a)   (b)

Figure 6. The stress-strain curve for bending testing specimens of PETG (a) 30mm/s (b) 25mm/s.

3.2 Bending testing

The results of the bending testing of PETG was demonstrated in Figure 6. Figures 6(a) and 6(b) illustrate the stress-strain curve of the specimens processed in 30mm/s and 25mm/s, respectively. The specimen specifications are depicted in Figure 7. Figure 6(a) shows that the trend of stress-stain curve of the 225°C and 245°C specimens are similar to each other, when the processed speed is set at 35mm/s. The maximized value of UBS and Young's modulus appears as a temperature of 235°C. The value of the UBS and Young's modulus specimens (25mm/s) is flat in the range of 225°～235°C, but it increases by 25% and 31%, respectively, when the temperature reaches 245°C. The results indicate that the higher temperature enhances the bending strength by minimizing the cave structure of the PETG material and the efficiency of printing decreases when the temperature is set under 230°C [8].

Figure 8 shows the value of Young's modulus of the bending test on PETG, and the results demonstrate that the higher temperature resulting from the lower value, shows that the temperature should be set as 235°C of the PETG specimens (30mm/s) to obtain the optimal value of Young's modulus. The trend of Young's Modulus of PETG specimens (25mm/s) increased by 36%, from 1.21Kpa to 1.64Kpa, at a temperature of

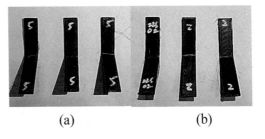

(a)   (b)

Figure 7. The specimens for bending testing specimens of PETG (a) 30mm/s, (b) 25mm/s.

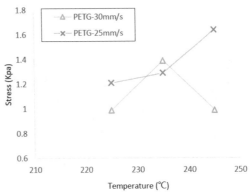

Figure 8. Young's modulus for the bending testing of PETG specimens.

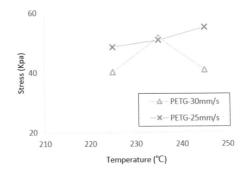

Figure 9. The UBS value of the bend testing specimens of PETG.

225°～245°C. In conclusion, the process that is based on FDM will reduce the mechanical properties of PETG.

Figure 9 shows that the temperature should be set at 235°C of the PETG specimens (30mm/s) to obtain the optimal value of the UBS. The average value of the UBS of the PETG specimens (25mm/s) is 44N, which is in the range of 225°～235°C. When the temperature reaches 245°C, the value of UBS increases by 25%, to 55.1N.

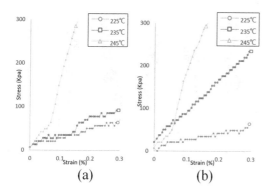

(a)   (b)

Figure 10. The stress-strain curve for compression testing specimens of PETG (a) 30mm/s (b) 25mm/s.

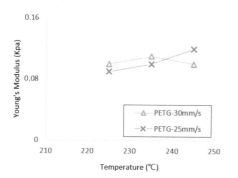

(a)   (b)

Figure 11. The specimens for compression testing of PETG (a) 30mm/s (b) 25mm/s.

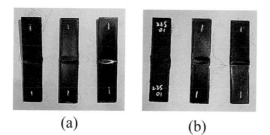

Figure 12. Young's modulus for the compression testing of PETG.

3.3 Compression testing

Figures 10(a) and 10(b) illustrate the stress-strain curve of the specimens that are processed in 30mm/s and 25mm/s, respectively. The specifications of the specimens are shown in Figure 11. It can be seen that the compression strength and Young's modulus increased with the temperature. The optimal value appears at 245°C of the nozzle temperature and 25mm/s of the processed speed. The results indicate that the condition of the cave structures that is caused by higher temperature is rare in PETG material, especially when the process speed is set at 25mm/s. The higher stickiness between each layer will result in the higher mechanical properties of PETG.

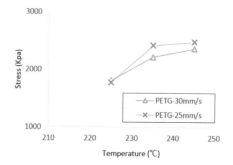

Figure 13. The compression strength of PETG.

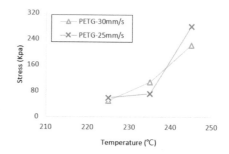

Figure 14. The yield strength for compression testing of PETG.

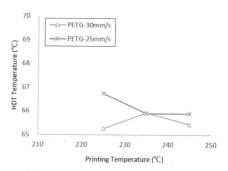

Figure 15. The value of HDT of PETG specimens.

Figure 12 illustrates the compression value of Young's modulus of PETG. It can be seen that the trend of Young's modulus increased when the process speed was set at 25mm/s. The value of compression strength of PETG is demonstrated in Figure 13. The PETG specimens processed in 30mm/s and 25mm/s has a stronger stickiness between each layer, with the increasing temperature. Figure 14 shows the value of the yield strength of compression testing. The value of the yield strength of PETG specimens apparently increased with the temperature. The maximized value of the yield strength of PETG specimens processed in 30mm/s and 25mm/s appears at 245°C.

3.4 Thermal deformation test

The obtained data from the thermal deformation test are shown in Figure 15. The results also indicated that

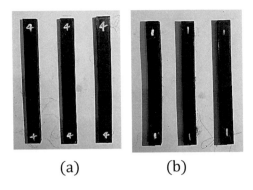

(a)   (b)

Figure 16.   The specimens for thermal deformation.

the PETG specimens (25mm/s) had a higher heat resistant temperature than the other specimens (30mm/s). The average value of the HDT of PETG specimens (25mm/s) was 66.18°C. In conclusion, PETG performed well on the mechanical properties of heat resistance. The specifications of the thermal deformation testing specimens of PETG are depicted in Figure 16.

4   CONCLUSION

The purpose of this research was to improve precision and to optimize the mechanical characteristics by adjusting the printing parameters of the temperature and the process speed, by using slice software that is based on FDM. The obtained data from tensile testing, compressive testing and thermal deformation testing indicated the following results:

(1) An increase in the process temperature of PETG could improve the precision value of the printing object, and it could apparently enhance the value of the tensile strength, compressive strength and bending strength.
(2) PETG specimens perform well on the tensile strength and bending strength, with an increasing temperature. The optimal value appears at 25mm/s; the value of Young's modulus has no significant diversification that is linked to temperature.

(3) The value of the compressive strength of PETG is out of proportion. The results showed that the properties of the material will change by using FDM technology, when the infill percentage is less than 100%.
(4) The PETG specimens show a high value of the distorted thermal temperature with a lower nozzle temperature. The process speed should be set at 40mm/s to obtain optimal results.

REFERENCES

Ahn, S. H., Montero, M., Odell, D., Roundy, S. & Wright, P. K. (2002). Anisotropic material properties of fused deposition modeling ABS. *Rapid Prototyping Journal*, 8(4), 248–257.

Ebel, E. & Sinnemann, T. (2014). Fabrication of FDM 3D objects with ABS and PLA and determination of their mechanical properties. *RTejournal*, 2014.

Kaveh, M., Badrossamay, M., Foroozmehr, E. & Etefagh, A. H. (2015). Optimization of the printing parameters affecting dimensionalaccuracy and internal cavity for HIPS material used in fuseddeposition modeling processes. *Journal of Materials Processing Technology*, 226, 280–286.

Serra, T., Ortiz-Hernandez, M., Engel, E., Planell, J. & Navarro, M. (2014). Relevance of PEG in PLA-based blends for tissue engineering 3D-printed scaffolds. *Materials Science and Engineering: C*, 38, 55–62.

Valerga, A., Batista, M., Salguero, J. & Girot, F. (2018). Influence of PLA Filament Conditions on Characteristics of FDM Parts. *Materials*, 11(8), 1322.

Jiang, D. & Smith, D. (2017). Anisotropic mechanical properties of oriented carbon fiber filled polymer composites produced with fused filament fabrication. *Additive Manufacturing*, 18, 84–94.

Santana, L., Alves, J., Sabino Netto, A. & Merlini, C. (2018). Estudo comparativo entre PETG e PLA para Impressão 3D através de caracterização térmica, *química e mecânica. Matéria (Rio de Janeiro)*, 23(4).

Guesssma, S., Belhabib, S. & Nouri, H. (2019). Printability and Tensile Performance of 3D Printed Polyethylene Terephthalate Glycol Using Fused Deposition Modelling. *Polymers*, 11(7), 1220.

Author index

Smart Science, Design and Technology

The main goal of this series is to publish research papers in the application of "Smart Science, Design & Technology". The ultimate aim is to discover new scientific knowledge relevant to IT-based intelligent mechatronic systems, engineering and design innovations. We would like to invite investigators who are interested in mechatronics and information technology to contribute their original research articles to these books.

Mechatronic and information technology, in their broadest sense, are both academic and practical engineering fields that involve mechanical, electrical and computer engineering through the use of scientific principles and information technology. Technological innovation includes IT-based intelligent mechanical systems, mechanics and systems design, which implant intelligence to machine systems, giving rise to the new areas of machine learning and artificial intelligence.

ISSN : 2640-5504
eISSN : 2640-5512

1. Engineering Innovation and Design: Proceedings of the 7th International Conference on Innovation, Communication and Engineering (ICICE 2018), November 9–14, 2018, Hangzhou, China

   Edited by Artde Donald Kin-Tak Lam, Stephen D. Prior, Siu-Tsen Shen, Sheng-Joue Young & Liang-Wen Ji

   ISBN: 978-0-367-02959-3 (Hbk + multimedia device)
   ISBN: 978-0-429-01977-7 (eBook)
   DOI: https://doi.org/10.1201/9780429019777

2. Smart Science, Design & Technology: Proceedings of the 5th International Conference on Applied System Innovation (ICASI 2019), April 12–18, 2019, Fukuoka, Japan

   Edited by Artde Donald Kin-Tak Lam, Stephen D. Prior, Siu-Tsen Shen, Sheng-Joue Young & Liang-Wen Ji

   ISBN: 978-0-367-17867-3 (Hbk)
   ISBN: 978-0-429-05812-7 (eBook)
   DOI: https://doi.org/10.1201/9780429058127

3. Innovation in Design, Communication and Engineering: Proceedings of the 8th Asian Conference on Innovation, Communication and Engineering (ACICE 2019), October 25–30, 2019, Zhengzhou, P.R. China

   Edited by Artde Donald Kin-Tak Lam, Stephen D. Prior, Siu-Tsen Shen, Sheng-Joue Young & Liang-Wen Ji

   ISBN: 978-0-367-17777-5 (Hbk)
   ISBN: 978-0-429-05766-3 (eBook)
   DOI: https://doi.org/10.1201/9780429057663